ENERGY SCIENCE AND ENGINEERING

AUTHOR(s)

M. Rizwan
Professor
Department of Electrical Engineering,
Delhi Technological University,
Delhi, INDIA

Majid Jamil
Professor
Department of Electrical Engineering,
Jamia Millia Islamia,
New Delhi, INDIA

KHANNA BOOK PUBLISHING CO. (P) Ltd.
4C/4344, Ansari Road, Darya Ganj,
New Delhi, India

AICTE

All India Council for Technical Education
Nelson Mandela Marg, Vasant Kunj,
New Delhi, India

Dear Readers,

To prevent the piracy, this book is secured with HIGH SECURITY HOLOGRAM on the front title cover. In case you don't find the hologram on the front cover title, please write us to at contact@khannabooks.com or whatsapp us at +91-99109 09320 and avail special gift voucher for yourself.

Specimen of Hologram on front Cover title:

Moreover, there is a SPECIAL DISCOUNT COUPON for you with EVERY HOLOGRAM.

How to avail this SPECIAL DISCOUNT:

Step 1: Scratch the hologram

Step 2: Under the scratch area, your "coupon code" is available

Step 3: Logon to www.khannabooks.com

Step 4: Use your "coupon code" in the shopping cart and get your copy at a special discount

Step 5: Enjoy your reading!

ISBN: 978-93-5538-514-7
Book Code: UG404EN

Energy Science and Engineering
by M. Rizwan, Majid Jamil

First Edition: 2026

Originally Published by:
All India Council for Technical Education, Ministry of Education, GOI

Exclusively Published & Distributed by:
Khanna Book Publishing Co. (P) Ltd.
4C/4344, Ansari Road, Daryaganj,
New Delhi (INDIA)

Visit us at: **www.khannabooks.com**
Write us at: **contact@khannabooks.com**

To view complete list of books, Please scan the QR Code: KPH

Printed in Bharat.

प्रो. म. जगदीश कुमार
अध्यक्ष
Prof. M. Jagadesh Kumar
Chairman

सत्यमेव जयते

आज़ादी का अमृत महोत्सव

अखिल भारतीय तकनीकी शिक्षा परिषद्
(भारत सरकार का एक सांविधिक निकाय)
(शिक्षा मंत्रालय, भारत सरकार)
नेल्सन मंडेला मार्ग, वसंत कुंज, नई दिल्ली–110070
दूरभाष : 011–26131498
ई–मेल : chairman@aicte-india.org

ALL INDIA COUNCIL FOR TECHNICAL EDUCATION
(A STATUTORY BODY OF THE GOVT. OF INDIA)
(Ministry of Education, Govt. of India)
Nelson Mandela Marg, Vasant Kunj, New Delhi-110070
Phone : 011-26131498
E-mail : chairman@aicte-india.org

FOREWORD

Engineers are the backbone of the modern society. It is through them that engineering marvels have happened and improved quality of life across the world. They have driven humanity towards greater heights in a more evolved and unprecedented manner.

The All India Council for Technical Education (AICTE), led from the front and assisted students, faculty & institutions in every possible manner towards the strengthening of the technical education in the country. AICTE is always working towards promoting quality Technical Education to make India a modern developed nation with the integration of modern knowledge & traditional knowledge for the welfare of mankind.

An array of initiatives have been taken by AICTE in last decade which have been accelerate now by the National Education Policy (NEP)2020. The implementation of NEP under the visionary leadership of Hon'ble Prime Minister of India envisages the provision for education in regional languages to all, thereby ensuring that every graduate becomes competent enough and is in a position to contribute towards the national growth and development through innovation & entrepreneurship.

One of the spheres where AICTE had been relentlessly working since 2021-22 is providing high quality books prepared and translated by eminent educators in various Indian languages to its engineering students at Under Graduate & Diploma level. For the second year students, AICTE has identified 87 books at Under Graduate and Diploma Level courses, for translation in 12 Indian languages - Hindi, Tamil, Gujarati, Odia, Bengali, Kannada, Urdu, Punjabi, Telugu, Marathi, Assamese & Malayalam. In addition to the English medium, the 1044 books in different Indian Languages are going to support to engineering students to learn in their mother tongue. Currently, there are 39 institutions in 11 states offering courses in Indian languages in 7 disciplines like Biomedical Engineering, Civil Engineering, Computer Science & Engineering, Electrical Engineering, Electronics & Communication Engineering, Information Technology Engineering & Mechanical Engineering, Architecture, and Interior Designing. This will become possible due to active involvement and support of universities/institutions in different states.

On behalf of AICTE, I express sincere gratitude to all distinguished authors, reviewers and translators from different IITs, NITs and other institutions for their admirable contribution in a very short span of time.

AICTE is confident that these out comes based books with their rich content will help technical students master the subjects with factor comprehension and greater ease.

M Jagadesh Kumar

(Prof. M. Jagadesh Kumar)

Acknowledgment

The authors are grateful to the authorities of AICTE, particularly Prof. M.Jagadesh Kumar, Chairman; Prof. M. P. Poonia, Vice-Chairman; Prof. Rajive Kumar, Member-Secretary and Dr. Amit Kumar Srivastava, Director, Faculty Development Cell for their planning to publish the books on "Energy Science and Engineering." We sincerely acknowledge the valuable contributions of the reviewer of the book Prof. Santanu Bandyopadhyay, Department of Energy Science and Engineering, Indian Institute of Technology Bombay for providing valuable inputs in all units of the book.

We would also like to extend our thanks to Prof. J.P. Saini, Vice Chancellor, Delhi Technological University, Prof. Madhusudan Singh, Registrar, Prof. Pragati Kumar, Head, Department of Electrical Engineering, Delhi Technological University, Delhi, Jamia Administration, Prof. Munna Khan, Head, Department of Electrical Engineering, JMI, Prof. Moinuddin, Prof. Sirajuddin Ahmad, and our Ph.D. students Astitva Kumar, Mohd. Bilal, Abdul Azeem, and Shamshad Ali.

This book is an outcome of various suggestions of AICTE members, experts, and authors who shared their opinion and thought to further develop engineering education in our country. Acknowledgments are due to the contributors and different workers in this field whose published books, review articles, papers, photographs, footnotes, references, and other valuable information enriched us at the time of writing the book.

M. Rizwan

Majid Jamil

Preface

The book titled "**Energy Science and Engineering**" is an outcome of the rich experience of two authors and a reviewer in the field of electrical engineering and energy sciences. The initiation of writing this book is to expose the basics of energy science to engineering students and enable them to get an insight into the subject. Keeping in mind the purpose of wide coverage as well as to provide essential supplementary information, we have included the topics recommended by AICTE in a very systematic and orderly manner throughout the book. Efforts have been made to explain the fundamental concepts of the subject in the simplest possible way.

During the process of preparation of the manuscript, we have considered the various standard text books, recent papers published in the field apart from the latest information available on relevant websites, and accordingly, we have developed contents in different units. While preparing the different sections, emphasis has also been laid on definitions and laws and also on comprehensive synopsis of formulae for a quick revision of the basic principles. The book covers fundamental as well as advanced knowledge in a very logical and systematic manner.

The objective of this book is to provide an introduction to energy systems and renewable energy resources, with a scientific examination of the energy field with an emphasis on alternative energy sources, their technology and application. This book will provide an opportunity to explore society's present energy needs and future energy demands, and examine conventional energy sources and systems, including fossil fuels and nuclear energy. More focus of this book is on renewable energy sources, sustainability & the environment.

Clean energy technologies and their importance in sustainable development, carbon footprint, energy and environment, trade and research policy, future energy use that can be influenced by economic, linkage between economic and environment outcomes are discussed by including the latest available statistical data. This book also covers topics on civil project development for creating energy infrastructure. Concepts of green building and green architecture, LEED ratings, energy auditing, and energy enterprises with some practical examples are presented in easy language to understand the fundamental concepts. It is important to note that in all the units, we have provided a QR Code to collect additional knowledge on the specific topic.

As far as the present book is concerned, "**Energy Science and Engineering**" is meant to provide a thorough grounding in energy and environment on the topics covered. This book will prepare engineering students to apply the knowledge to meet the challenges of protecting the environment and providing solutions for the development of sustainable, environment-friendly clean and green energy to tackle 21st century and onward engineering challenges and address the related aroused questions. The subject matters are presented in a constructive manner so that an Engineering degree prepares students to work in different sectors or in national laboratories at the very forefront of technology.

We sincerely hope that the book will inspire the students to learn and discuss the ideas behind the sustainable development of energy and will surely contribute to the development of a solid foundation of the subject. We would be thankful to all beneficial comments and suggestions which will contribute to the improvement of future editions of the book. It gives us immense pleasure to place this book in the hands of the teachers and students. It was indeed a big pleasure to work on different aspects covered in the book.

M. Rizwan

Majid Jamil

Outcome Based Education

For the implementation of an outcome-based education the first requirement is to develop an outcome-based curriculum and incorporate an outcome-based assessment in the education system. By going through outcome-based assessments evaluators will be able to evaluate whether the students have achieved the outlined standard, specific and measurable outcomes. With the proper incorporation of outcome-based education there will be a definite commitment to achieve a minimum standard for all learners without giving up at any level. At the end of the program running with the aid of outcome-based education, a student will be able to arrive at the following outcomes:

PO1. Engineering knowledge: Apply the knowledge of mathematics, science, engineering fundamentals, and an engineering specialization to the solution of complex engineering problems.

PO2. Problem analysis: Identify, formulate, review research literature, and analyze complex engineering problems reaching substantiated conclusions using the first principles of mathematics, natural sciences, and engineering sciences.

PO3. Design/development of solutions: Design solutions for complex engineering problems and design system components or processes that meet the specified needs with appropriate consideration for public health and safety, and cultural, societal, and environmental considerations.

PO4. Conduct investigations of complex problems: Use research-based knowledge and research methods including design of experiments, analysis and interpretation of data, and synthesis of the information to provide valid conclusions.

PO5. Modern tool usage: Create, select, and apply appropriate techniques, resources, and modern engineering and IT tools including prediction and modeling to complex engineering activities with an understanding of the limitations.

PO6. The engineer and society: Apply reasoning informed by the contextual knowledge to assess societal, health, safety, legal and cultural issues, and the consequent responsibilities relevant to the professional engineering practice.

PO7. Environment and sustainability: Understand the impact of professional engineering solutions in societal and environmental contexts, and demonstrate the knowledge of, and need for sustainable development.

PO8. Ethics: Apply ethical principles and commit to professional ethics and responsibilities and norms of the engineering practice.

PO9. Individual and team work: Function effectively as an individual, and as a member or leader in diverse teams, and in multidisciplinary settings.

PO10. Communication: Communicate effectively on complex engineering activities with the engineering community and with society at large, such as being able to comprehend and write effective reports and design documentation, make effective presentations, and give and receive clear instructions.

PO11. Project management and finance: Demonstrate knowledge and understanding of the engineering and management principles and apply these to one's own work, as a member and leader in a team, to manage projects and in multidisciplinary environments.

PO12. Life-long learning: Recognize the need for, and have the preparation and ability to engage in independent and life-long learning in the broadest context of technological change.

Course Outcomes

Upon successful completion of the course, the students will be able to:

CO-1: Understand the fundamental aspects of energy and impacts of energy on society, environment & climate change.

CO-2: Have the basic understanding of scientific concepts of energy, energy sources and energy storage.

CO-3: Understand the basics of clean energy technologies and how the use of energy influence economic, trade and environment.

CO-4: Understand the civil construction required to set up different types of underground, on and off shore power plants.

CO-5: Realize the safety aspects of nuclear fuel & its disposal and construction of nuclear power plants

CO-6: Have an idea about green buildings, energy auditing and energy conservation.

Course Outcomes	Expected Mapping with Programme Outcomes (1- Weak Correlation; 2- Medium correlation; 3- Strong Correlation)											
	PO-1	PO-2	PO-3	PO-4	PO-5	PO-6	PO-7	PO-8	PO-9	PO-10	PO-11	PO-12
CO-1	3	3	3	2	-	3	3	3	3	3	3	3
CO-2	3	3	2	1	-	3	2	3	1	3	3	3
CO-3	3	3	3	1	2	3	3	3	3	3	3	3
CO-4	3	2	1	1	2	1	1	-	-	1	2	1
CO-5	3	1	1	-	-	1`	1	1	2	1	3	1
CO-6	3	2	2	-	1	3	3	3	3	3	3	3

Guidelines for Teachers

To implement Outcome Based Education (OBE) knowledge level and skill set of the students should be enhanced. Teachers should take a major responsibility for the proper implementation of OBE. Some of the responsibilities (not limited to) for the teachers in OBE system may be as follows:

- Within reasonable constraint, they should manoeuvre time to the best advantage of all students.
- They should assess the students only upon certain defined criterion without considering any other potential ineligibility to discriminate them.
- They should try to grow the learning abilities of the students to a certain level before they leave the institute.
- They should try to ensure that all the students are equipped with the quality knowledge as well as competence after they finish their education.
- They should always encourage the students to develop their ultimate performance capabilities.
- They should facilitate and encourage group work and team work to consolidate newer approach.
- They should follow Blooms taxonomy in every part of the assessment.

Bloom's Taxonomy

Level	Teacher should Check	Student should be able to	Possible Mode of Assessment
Create	Students ability to create	Design or Create	Mini project
Evaluate	Students ability to justify	Argue or Defend	Assignment
Analyze	Students ability to distinguish	Differentiate or Distinguish	Project/Lab Methodology
Apply	Students ability to use information	Operate or Demonstrate	Technical Presentation/ Demonstration
Understand	Students ability to explain the ideas	Explain or Classify	Presentation/Seminar
Remember	Students ability to recall (or remember)	Define or Recall	Quiz

Guidelines for Students

Students should take equal responsibility for implementing the OBE. Some of the responsibilities (not limited to) for the students in OBE system are as follows:

- Students should be well aware of each UO before the start of a unit in each and every course.
- Students should be well aware of each CO before the start of the course.
- Students should be well aware of each PO before the start of the program.
- Students should think critically and reasonably with proper reflection and action.
- Learning of the students should be connected and integrated with practical and real life consequences.
- Students should be well aware of their competency at every level of OBE.

Abbreviations and Symbols

List of Abbreviations

General Terms			
Abbreviations	**Full form**	**Abbreviations**	**Full form**
AERB	Atomic Energy Regulatory Board	LCA	Life Cycle Analysis
BWR	Boiling Water Reactor	LCOE	Levelized Cost of Energy
CFC	Chlorofluorocarbons	LCOS	Levelized Cost of Storage
CFLS	Compact Fluorescent Lamps	LEED	Leadership in Energy and Environmental Design
CT	Cooling Tower	LED	Light Emitting Diode
DEFRA	Department Of Food and Rural Affair	LWGR	Light Water Graphite Reactor
DW	Dry Well	MNRE	Ministry of New and Renewable Energy
EC	Energy Conversion	NEP	National Electricity Plan
EE	Energy Efficiency	NPP	Nuclear Power Plant
EKC	Empirical Kuznet Curve	NPCIL	Nuclear Power Corporation of India Limited
EPA	Environment Protection Agency	PET	polyethylene terephthalate
eQUEST	Energy Quick energy simulation	PSAT	Pumping System Assessment Tool
EROEI	Energy Returned on Energy Invested	PWR	Pressurized Water Reactor
ERV	Energy Recovery Ventilation	RES	Renewable Energy Sources
FC	Fuel Cell	SECI	Solar Energy Corporation of India
FSAR	Final Safety Analysis Report	SMES	Superconducting Magnetic Energy Storage
FPSO	Floating Production Systems	SPeAR	Sustainable Project Appraisal Routine
GHG	Greenhouse gases	SPV	Solar Photovoltaic
GoI	Government of India	SSAT	Steam System Assessment Tool

General Terms			
Abbreviations	**Full form**	**Abbreviations**	**Full form**
IAEA	International Atomic Energy Agency	TLP	Tension-Leg Platform
IGBC	Indian Green Building Council	UPS	Uninterruptible Power Supply
IHAT	Indoor Humidity Assessment Tool	USGBC	United States Green Building Council
IPCC	Intergovernmental Panel on Climate Change	VALCOE	Value Adjusted Levelized cost of Electricity
LACE	Levelized Avoided Cost of Electricity	WRI	World Resource Institute

List of Figures

Unit 1 Introduction to Energy Science

Unit 2 Energy Sources

Unit 3 Energy and Environment

Unit 4 Civil Engineering Projects Connected with Energy Sources

List of Tables

CONTENTS

1 Introduction to Energy Science

UNIT SPECIFICS

In this unit, the following topics have been discussed for basic understating related to energy science:

- Concept of energy
- Impacts of energy on society, environment, and climate change
- Share of renewable energy resources in the power sector
- Promotion of renewable energy-based technologies by the Government

Here, in this unit, the main focus is to aware the academicians and students of essential principles and fundamental principles for energy education which introduces energy concepts, energy utilization, and energy decisions. Energy is an issue that is intrinsically multidisciplinary in nature. The fundamental concepts in energy are found in almost all, if not all, academic areas. This handbook is meant to be utilised by people from many fields. An integrated and systems-based approach to energy knowledge is strongly recommended.

Energy has played a critical role in human civilisation, and the use of energy sources, notably fossil fuels, has drastically affected people's living conditions. The spectrum of human applications for various energy sources has grown to the point that contemporary living would be nearly impossible without them. To put it another way, human life has become increasingly reliant on energy consumption, yet the advancements it allows come at a high cost to the environment and vulnerable communities. Energy may be compared to a driving force that propels a variety of desired outcomes, including personal comfort and well-being, social welfare, economic growth and prosperity, and technical, industrial, and sustainable development.

RATIONALE

This unit introduces fundamental concepts of energy sciences and historical interpretations in the context of energy. Different forms of energy and their scientific developments, sources of energy, and the impact of energy on society and the environment are described here. Merits and demerits of various renewable energy sources and the present trend of development of sustainable energy in developing and developed countries are also projected in this unit.

PRE-REQUISITES

Basic Knowledge of Physics-XII standard

UNIT OUTCOMES

List of outcomes of this unit is as follows

U1-O1: To know the historical interpretation and scientific principles of energy.

U1-O2: To understand the concept of force, energy, and power.

U1-O3: To realize the impacts of energy on the environment and related climatic issues.

U1-O4: To know the share of renewable energy at the national and global levels.

U1-O5: To understand the initiative and policies of the Government related to energy resources, environment, and climate change.

Unit outcomes	EXPECTED MAPPING WITH COURSE OUTCOMES (1- Weak Correlation; 2- Medium correlation; 3- Strong Correlation)					
	CO-1	CO-2	CO-3	CO-4	CO-5	CO-6
U1-O1	3	3	1	-	-	-
U1-O2	3	2	1	-	-	-
U1-O3	3	3	3	-	-	-
U1-O4	1	-	1	-	-	-
U1-O5	-	-	3	-	-	1

1.1. Historical Interpretation and Scientific Principles

1.1.1. Concept of Energy in Ancient Indian Text

The Indian civilisation, one of the world's oldest civilizations, has a long history of science and technology. India was a place of sages and seers as well as philosophers and scientists in ancient times. According to research, India actively contributed to the field of science and technology centuries before modern laboratories were set up, from creating the greatest steel in the world to teaching the world how to count. Many of the old Indian thoughts and methodologies have shaped and reinforced the foundations of modern science and technology. The concept of energy science is a part of classical physics, and Indian Vedas and Purana introduced Vedic Physics closely related to matter and energy [1].

In the ancient text, the method of producing electricity using an earthen pot, copper plate, copper sulfate, wet sawdust, and zinc amalgam can be found. This can be considered as the electric battery of ancient times. The above process is known as 'Agastya Samhita' inspired by the name of Sage Agastya. When a cell was prepared according to this process, it gave an open circuit voltage of 1.138 V and a short circuit current of 23 mA.

Please scan the QR code for further information on various topics related to 'Energy in Ancient Indian Text'.

Newton's three laws of motion, published in 1687, were already mentioned in Vaisheshika Philosophy. According to this literature, motions are of five types as shown in the Fig.1.1.

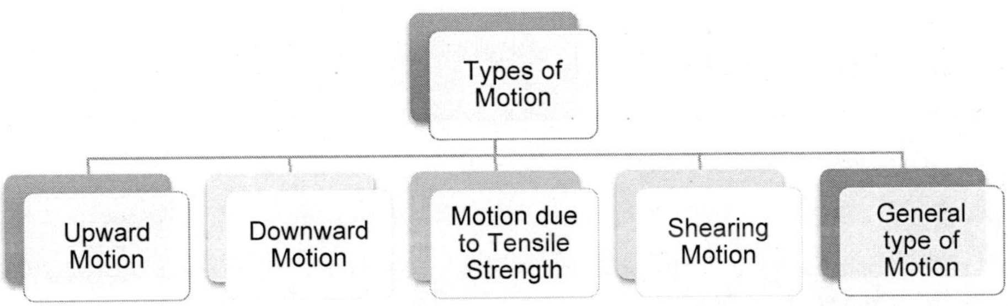

Fig. 1. 1. Different types of motion in the ancient Indian text

1.1.2. Historical Concept of Energy in Science and Its Measurements

The word 'energy' is derived from the Greek word 'energeia', which first appeared in the 4th century BCE works of Aristotle. 'Energeia' refers to as enaction that makes matter move and embodiment. Energy is further categorized as potential energy and kinetic energy. The 'force' gained by two falling bodies, one of mass m dropped a distance $4h$, and another of mass $4m$ dropped a distance h, led Leibniz to define vis viva. The rule of falling bodies was used by Leibniz to demonstrate that an item with a mass m has twice the velocity of an object with a mass $4m$ because it falls four times the distance. Because both bodies must be lifted with the same 'power,' the value of vis viva should be linked to mv^2 rather than mv. Furthermore, Leibniz claimed that mass multiplied by velocity squared is preserved [2].

Around the start of the 18th century, systematic analysis of thermal energy began under the auspices of Stahl's caloric theory that treated heat as a fluid ("phlogiston"), with no inkling of course about the relation to mechanical energy. This did not stop Newcomen from inventing a new steam engine in 1712 and Watt from improving it in 1763-1775 (he also introduced a unit of power, "horsepower"). After Lavoisier's work on oxygenation (1783), the phlogiston theory was modified into a caloric one, which remained in place until mid- 19th century. Carnot even developed the basics of the theory of heat engines based on it (1824). In 1807, Thomas Young was the first person to use the modern sense of the word 'energy' for the quantity of mass multiplied by velocity squared. William Thompson, who later became Lord Kelvin, is credited with introducing the concept of kinetic energy in 1849. We now associate the concept of an object's kinetic energy with the quantity of one-half of its mass multiplied by its velocity squared. Gustave Gaspard Coriolis, in 1829, described kinetic energy in its modern sense, and William Rankine coined the term potential energy in 1853.

Measurements of the mechanical equivalent of heat were first performed by Rumford in 1798. In the 1840s, Joule extended the work of Rumford; this further led to the formulation of general energy conservation law by Mayer in 1841. In 1850, Clausius reformulated Carnot's work on heat as energy, which led to the acceptance of heat as another form of energy. In 1851 Thomson already could write that "heat is not a substance, but a dynamical form of mechanical effect, we perceive that there must be an equivalence between mechanical work and heat, as between cause and effect". By the time the International System of Units (SI) and centimetres grams seconds (CGS) system of Units were undergoing a major overhaul of units in 1860-1880s, mechanical, thermal, and electric energies were treated uniformly. In the SI, energy is measured in joules. One joule is equal to the work done by a one-newton force acting over a one-meter distance.

1.1.3. Theory of Energy in Different Scientific Descriptions

Energy is described as the ability to perform work in physics. Potential, kinetic, thermal, electrical, chemical, radioactive, and other forms of energy exist [3]. These different forms of energy are briefly shown in Fig. 1.2. Heat and work (energy exchanged through one body to another) are also involved. Once the energy has been transferred, it's often classified as per its type. As an outcome, heat transferred could be turned into thermal energy, whilst effort performed could be translated to mechanical energy. Other techniques of changing energy from one form to another prevail. Many types of technologies, such as gasoline heat engines, generators, batteries, fuel cells, and magnetohydrodynamic systems, create useful electromechanical energy [4].

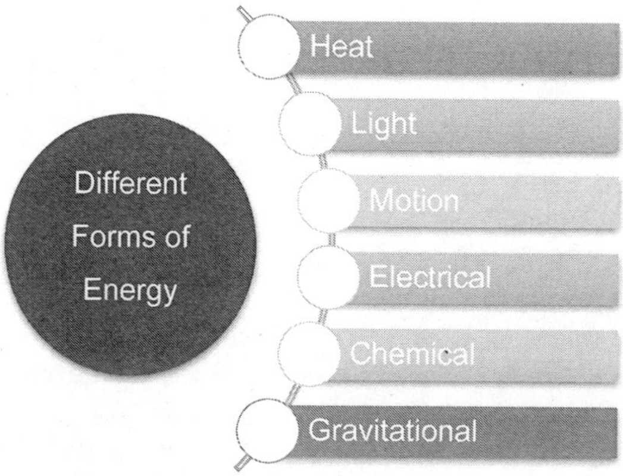

Fig. 1. 2. Different forms of energy

All forms of energy are related to motion. Kinetic energy is present when one body is in motion. A tensioned device, such as a bow or spring, has the ability to create motion even when it is at rest; its arrangement incorporates potential energy. Because nuclear energy is generated by the configuration of quantum particles inside an atom's nucleus, it is also known as potential energy.

Energy cannot be created or destroyed, only altered. This principle is known as the conservation of energy, sometimes known as the first rule of thermodynamics. When a box slides down a hill, the potential energy it has from being high on the slope is converted to kinetic energy or motion energy. When the box comes to a halt owing to friction, the kinetic energy from its movement is converted to thermal energy, which warms the box and the slope [5].

1.1.4. Force, Energy, and Power

Force

Push and pull are the basic principles of force that we have studied. Whenever force is exerted to an object, it affects its movement or structure.

Work

We consider any beneficial physical or mental labor to be work in our daily lives. Playing in a park, conversing with colleagues, singing a song, going to the movies, and attending a function are all examples of activities that are not often considered labor. What we consider to be "work" is determined by how we define it. In science, we use and define the term "labor" differently [6]. As a result, the work done by a force acting on an entity is equivalent to the force's magnitude multiplied by the distance traveled in the force's direction. Work has no direction and merely a magnitude.

$$W = F \times d \tag{1.1}$$

Energy

The capacity to do work is defined as Energy. Fig. 1.3 shows the various types of energy that can be measured. The SI unit of energy is joule (J). As a result, an object's energy is calculated

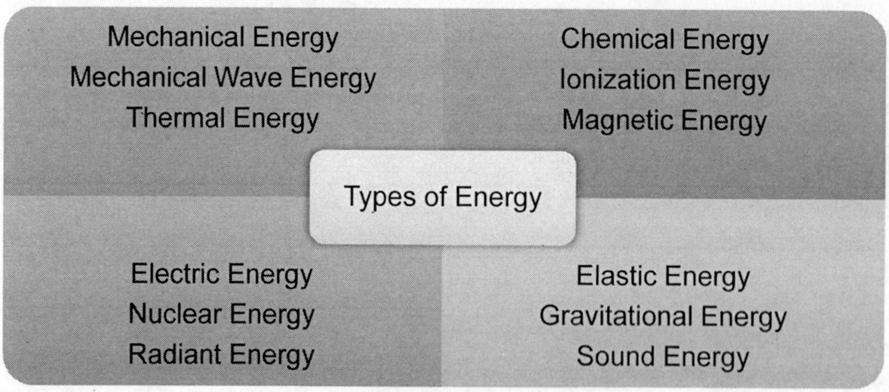

Fig. 1. 3. Different types of energy

based on its ability to perform work. 1 J is the amount of energy needed to perform one joule of work. The kilojoule (kJ) is a greater unit of energy that is sometimes used. 1000 J is equal to 1 kJ. The commercial unit of energy is watt-hour (Wh). The energy used in households, industries, and commercial establishments is usually expressed in kilowatt-hours. For example, electrical energy used during a month is expressed in terms of ‘units. Here, 1 ‘unit’ means 1 kilowatt-hour.

Power

Power is defined as the rate at which work is done. The SI unit of power is watt (W). 1 watt is the power of an object, which does work at the rate of 1 joule per second.

1.1.5. Law of Conservation of Energy

Energy cannot be generated or destroyed, according to the rule of conservation of energy. It can, however, be changed from one form to another. The overall energy of an isolated system remains constant when all kinds of energy are considered. Moreover, it can be further elaborated that energy remains constant for a closed system, i.e., isolated from its surroundings when all the energy sources are considered. Some of the examples of the conversion of energy are depicted in Fig. 1.4.

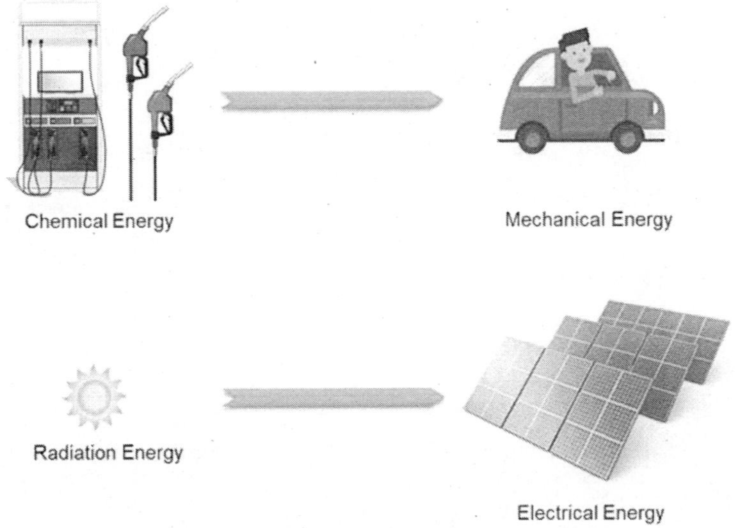

Fig. 1. 4. Examples of energy conversion and law of conservation

1.2. Impact of Energy on Society, Environment, and Climate

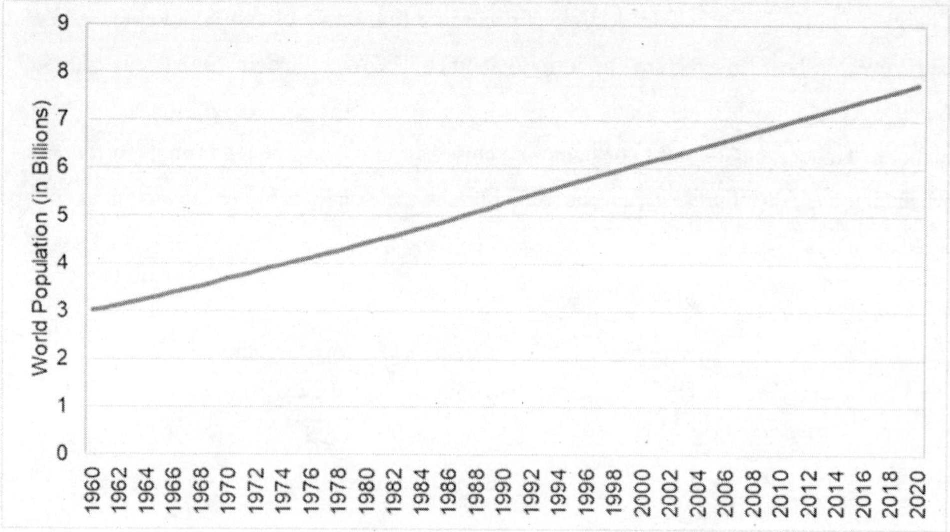

Fig. 1. 5. World's population since 1960

For more information regarding the population data, please scan the QR code

In the past six decades, the world's population has increased by twice than the population recorded in 1960, as shown in Fig. 1.5. Due to the rising demands of the ever-increasing population for a better quality of life, the consumption of energy is also on the rise. This has led to new scientific innovations and inventions that have impacted society and the environment in both ways, i.e., good and bad.

1.2.1. Impact of Energy on Society

The influence of energy on society is both beneficial and unfavourable. Humans benefit from having access to plentiful, inexpensive, safe, efficient, and clean energy. However, energy production, transmission, and consumption may have detrimental effects on a society's health, ecology, and economy. Furthermore, reliance on energy imports might put a country's security at risk. The consequences of energy expenditures are not the same for everyone. Because they have a lower ability for adaptation and negotiating leverage than wealthy civilizations, poor or disadvantaged cultures are more likely to face unfavourable effects of energy decisions. As a result, advances in energy accessibility, safety, or cost can help vulnerable people significantly.

Human society's energy usage is influenced by social status and access to technology. As a result, technological and societal change can help mankind utilize less energy [7]. Reduction in energy use does not always imply a worsening in quality of life. Energy conservation may improve the quality of life in many circumstances by reducing environmental hazards, increasing economic and national security, and saving money. The structure of civilization has been influenced by the availability of low-cost, reliable energy resources. The effects of fossil fuels, which are used to create energy, have further defined societal structure since the industrial revolution. Poor communities are typically willing to embrace inefficient energy installations because they provide much-needed jobs. The consumption of energy across the globe is depicted in Fig. 1.6.

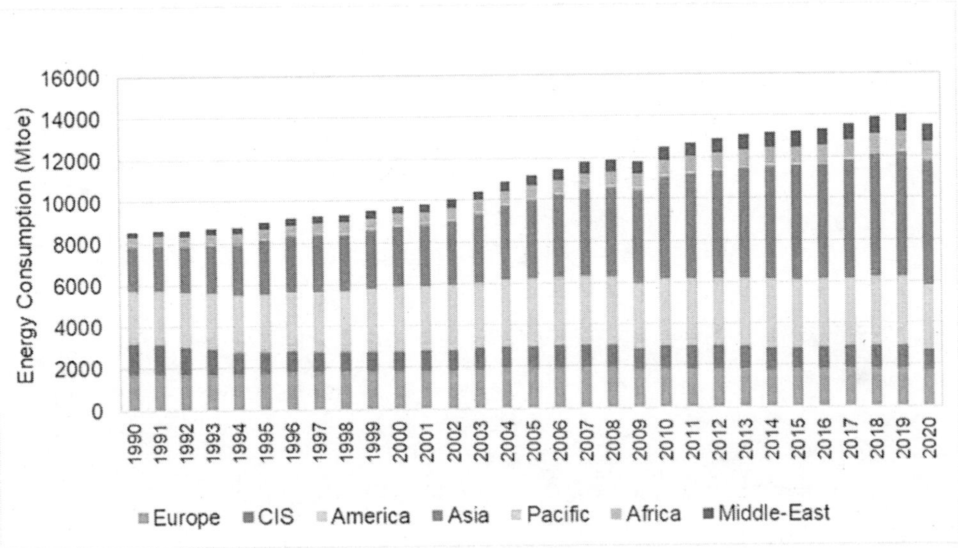

Fig. 1. 6. World's energy consumption since 1990

For the latest data on energy consumption, please scan the QR code

1.2.2. Impact of Energy on Environment and Climate

The energy consumption of the world has been on a constant rise in the last three decades barring the year 2020, due to unforeseen circumstances of lockdown and restrictions. This has led to more burning of fossil fuels which in turn have an adverse impact on the environment such as global warming, acid rain, higher emission of greenhouse gases, etc. Moreover, this higher rate of burning of conventional sources of energy (i.e., coal, oil, and petroleum) have led to drastic climatic changes. Increasing pollutants and ecosystem deterioration are all part of an ever-growing list of environmental issues that affect ever-larger regions. Energy supply and use issues are linked to environmental concerns such as air pollution, acid rainfall, ozone layer depletion, rainforest loss, and radioactive material emissions, in addition to global warming. If mankind wants to attain a brighter future economy with minimum environmental consequences, these concerns must be addressed concurrently. There is enough evidences to show that if people continue to degrade the environment, the future will suffer. The worldwide greenhouse gases (GHG) emissions have increased tremendously over the past few years, causing the three most pressing environmental challenges for the world: acid precipitation, stratospheric ozone depletion, and global climate change [9]. Fig. 1.7. represent the greenhouse gases emission trend of the world since 1990 till 2018.

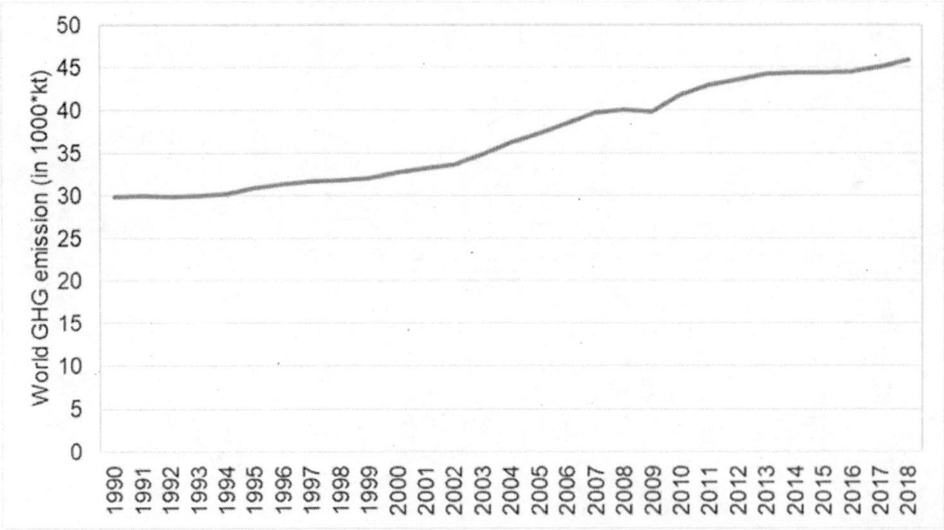

Fig. 1. 7. GHG emission of the world since 1990

For accessing the data regarding GHG emissions of the world, please scan the QR code

Renewable energy sources, energy conservation, and energy storage technologies, among others, have recently emerged as promising answers to the present environmental concerns connected with hazardous pollutants. The two most significant topics, sustainable energy technology and energy conservation, will be discussed. Several countries think that wind, solar, and other renewable energy sources are the key to a clean energy future, and they are working hard to be the world's leader in inventing, producing, and implementing these techniques.

1.3.Introduction to Energy Systems and Resources

1.3.1. Types of Energy Resources

Mainly, energy sources are classified into two broad categories, non-renewable and renewable energy resources. Non-renewable energy resources are conventional energy sources that cannot be replenished over a period of time and can get depleted. These sources are derived from fossils over a period of time under extreme pressure and temperature. Non-renewable energy resources are oil, natural gas, coal, and nuclear energy.

Renewable energy sources are the ones that can be replenished naturally over a period of time and can be utilized repeatedly. These sources include biomass energy (such as ethanol), hydropower, geothermal power, wind energy, and solar energy.

Some of the positives of renewable energy are:

- Sustainable and will not run out very soon
- Environment friendly and low GHG emission
- Factors such as trade laws, territorial claims, and political instabilities do not impact renewable energy sources
- Eventually leads to improved public health and happier lives
- Requires less maintenance
- Decrease the risk of energy crisis
- Reduce the risk of energy prices

Some of the disadvantages associated with renewable energy are:

- Affected by meteorological parameters thus not available round the clock
- Energy conversion efficiency of this technology is low
- Higher initial cost of investment
- Need to ample land to set up a renewable energy plant

1.3.2. Global Share of Conventional and Renewable Energy

The major energy consumption around the world is from non-renewable energy sources i.e., coal, oil and gases. The rest of the energy is consumed in the form of biomass and electrical energy around the world. China consumes almost half of the world's coal consumption, and is the only major user to see an increase in coal usage by 0.6%. Despite China's strategy of reducing coal usage in primary energy, the country's demand for coal has still not hit its peak.

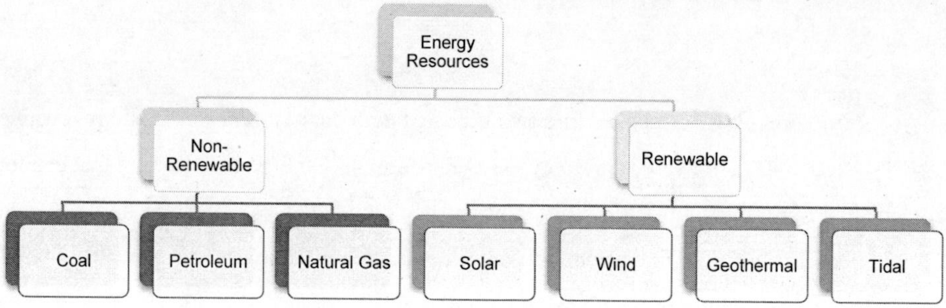

Fig. 1. 8. Types of energy sources

Whereas, in other nations, a drop in coal and other fossil fuels usage was observed due to government initiative, awareness among people and the pandemic. Due to public and corporate climate legislation, competition from cheaper gas-fired and renewable power generation, and coal-fired power plant shutdowns, notably in the United States, the crisis has escalated the declining trend in coal usage seen in previous years, as shown in Fig. 1.10.

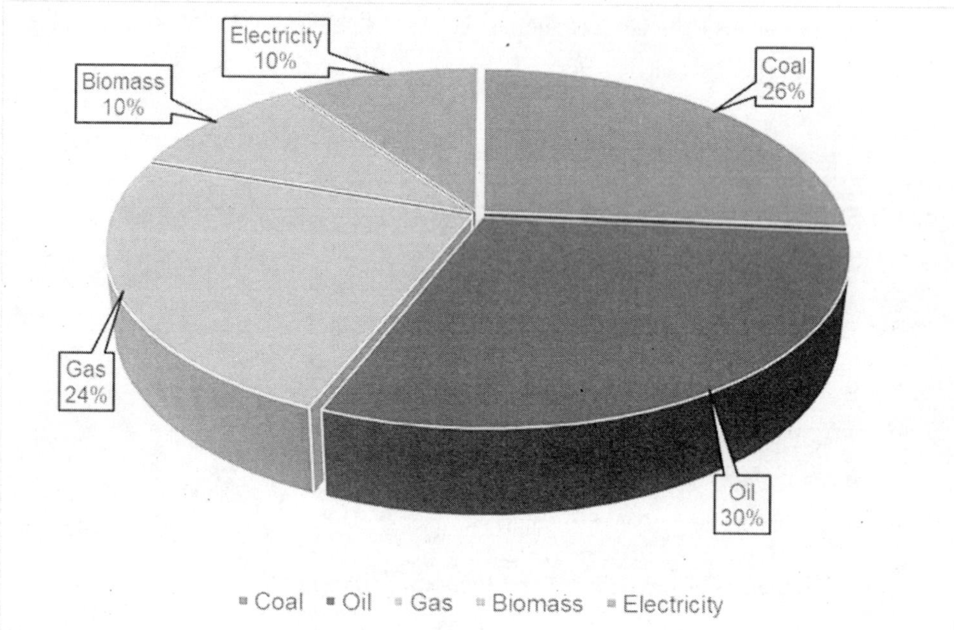

Fig. 1. 9. Sources used for energy consumption of the world

To access the data of Fig.1.9, please scan the QR code

From the worldwide trend, generation from renewables (including hydro) is observed to increase by over 6% in 2020, owing to continued expansion in wind and solar output, although hydropower's proportion of the global power mix has remained relatively consistent at around 16% since 2000. The EU, the US, China, India, Japan, Chile, and Australia are all experiencing a rise in renewable energy output, thanks to aggressive climate regulations and falling solar and wind technology costs. Following a rising trend that began in the 2000s, the share of renewable energy sources (including hydropower) in worldwide power generation is increased by approximately 1% to over 29 percent in 2021 [8].

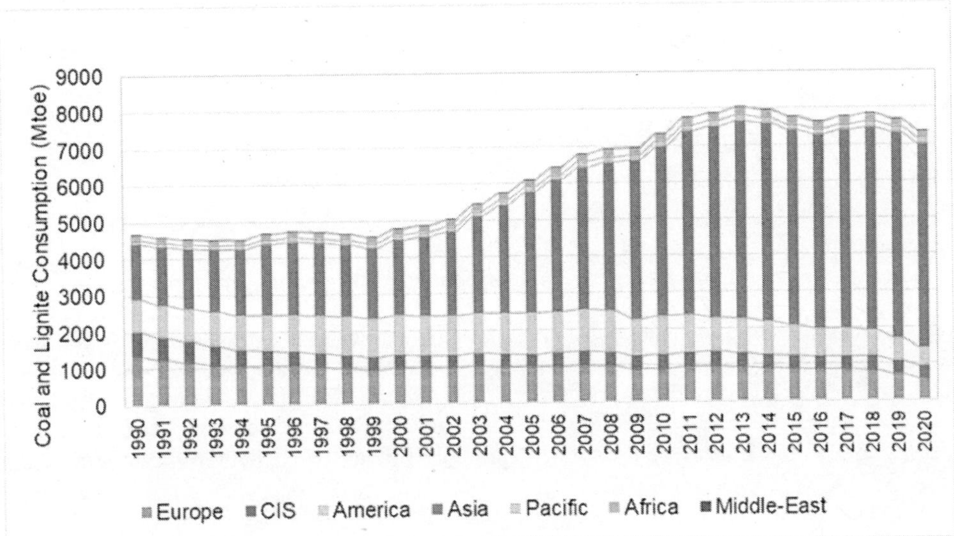

Fig. 1. 10. Coal and Lignite (major non-renewable source) for world's energy consumption

For more information regarding Fig. 1.10, please scan the QR code

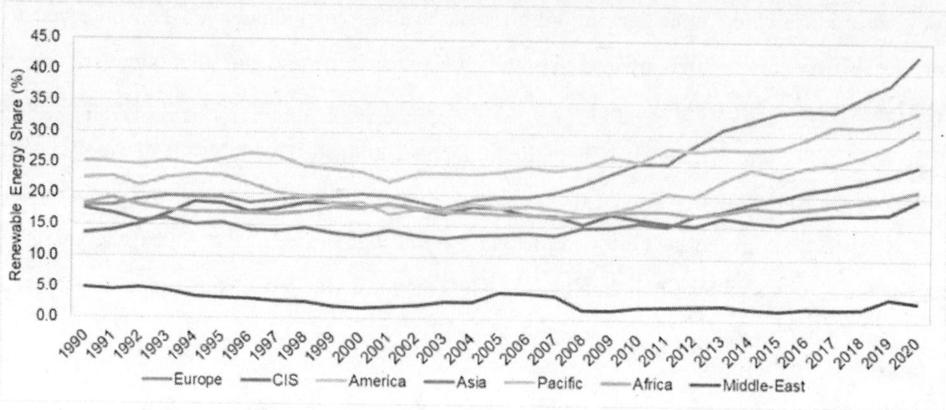

Fig. 1. 11. Renewable energy share of the world

For more information regarding renewable energy share of the world, please scan the QR code

1.3.3. National Share of Conventional and Renewable Energy

The growth in energy demand, ever-increasing population, over-exploitation of fossil fuels, and economic policies are the driving factors for RE in the Indian power sector. This led to the setting up of the Ministry of New and Renewable Energy (MNRE) in 1982 by the Indian government to promote and form policies regarding energy generation using RE sources. The total installed capacity of India considering central, state, and private sectors cumulatively is at 399.497 GW as of 31st March 2022. The RE capacity of India constitutes about 35.10% of the total installed capacity. Among the RE sources, SPV has the maximum installed capacity of 53.997 GW followed by wind power with 40.358 GW, this accounts to major chunk of RE in India at 13.5% for solar and 10.3% for wind. Small hydro and bio-powered energy are at 4.849 GW and 10.685 GW, respectively in India [10]. A glance of Indian power sector is presented in Table 1 and Fig. 1.12. By 2030 India aims for achieving 40% of power from renewables.

Table 1.1 India's power sector at a glance (GW)

Thermal			Nuclear	Renewable	
Coal	Gas	Diesel		Hydro	Other Renewables
210.700	24.900	0.51	6.780	46.723	156.608

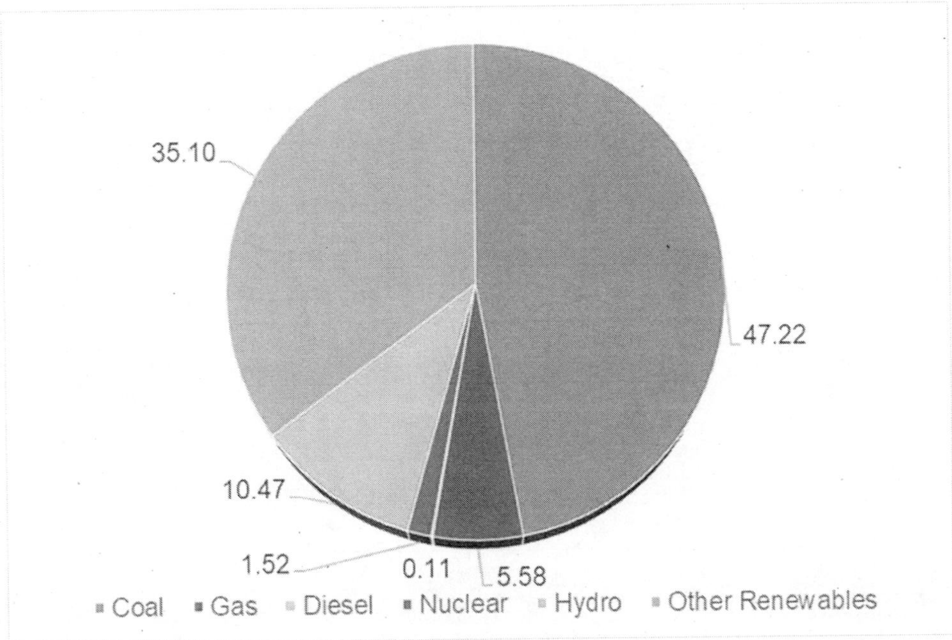

Fig. 1. 12. Indian power scenario

For latest information regarding Indian power scenario, please scan the QR code

The year-wise power generation growth in India is depicted, it illustrates the significant progress of India, in the field of the power sector. The progressive growth in the field of power, also explains the economic growth of the country.

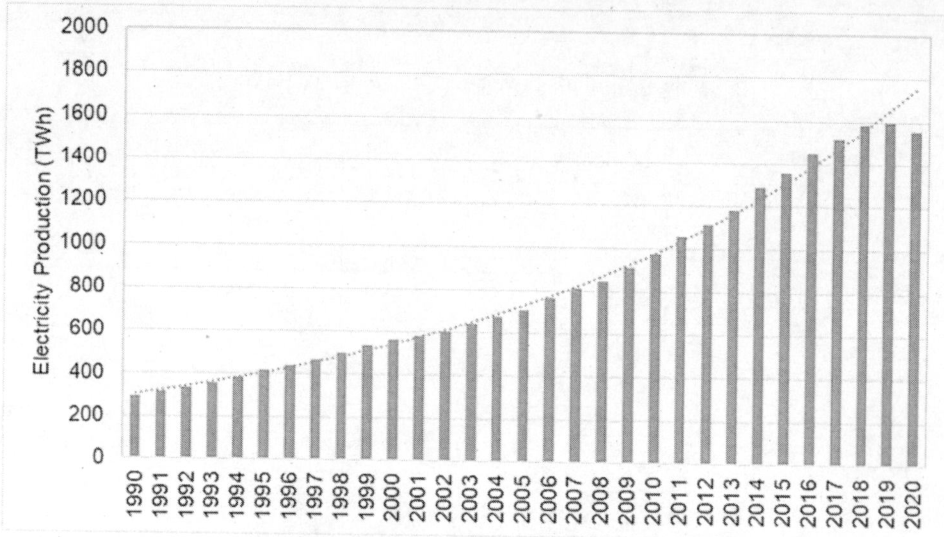

Fig. 1. 13. Electricity production of India since 1990

The electricity production data of India can be accessed by scanning the QR code

Table 1. 2 India's Renewable energy sector at a glance (MW)

Bio-Power	Waste to Energy	477
	Biomass	10206
Small Hydro		4849
Wind Power		40358
Solar Power		53997
Total		156608

For latest information regarding Indian renewable energy sector scan the QR code

1.4.Government of India Initiatives to Promote Renewable Energy

Government of India (GoI) in the recent years has taken major strides in formulating policies and schemes to promote renewable energy in India. The National Electricity Plan [NEP] framed by the Ministry of Power has proposed a ten-year action plan with detailed objectives to light up every home, even in the remotest of villages in India. The Government of India has also established Solar Energy Corporation of India (SECI), an autonomous body for the implementation of National Solar Mission, all across India effectively. The Government has also set up National Institute of Solar Energy (NISE), which is an independent institute for carrying out the latest research and development in the area of solar power energy and coordinating the training programs. Some of the schemes and initiatives taken up by GoI are [11-13]:

Table 1.3 Initiatives and Policies taken up by GoI for promoting RE

Biomass and Biogas		
Scheme/Policy	**Period**	**Salient Features**
Biomass based Cogeneration in Sugar Mills and other Industries	11.05.2018 to 31.03.2020	Financial Assistantship of 25 and 50 lacs for bagasse and non-bagasse cogeneration
Biogas Power Generation and Thermal Energy Programme (BPGTP)	01.04.2017 to 31.02.2021	Subsidy from Rs. 12,500/- to Rs. 40,000/- for setting up off-grid biogas-based power generation plant.
Solar Photovoltaic		
Development of Solar Parks	Upto 2022-23	The scheme envisages supporting the States/UTs in setting up solar parks at various locations in the country with a view to create required infrastructure for setting up of solar power projects.
Jawaharlal Nehru National Solar Mission Phase II	14.03.2016 to 31.03.2019	Funding of Rs. 5050 cr to be provided for installation of 5000MW grid connected solar photovoltaic projects.
Grid Connected Solar Rooftop Programme	Till 31.12.2022	Subsidy provided to residential/apartment owners for setting up of domestic solar photovoltaic on grid system.
Pradhan Mantri Kisan Urja Suraksha evam Utthaan Mahabhiyaan (PM-KUSUM)	Till 31.12.2022	Support to agriculture sector and farmers for utilizing and installing off-grid solar photovoltaic based water pumps.
Atal Jyoti Yojana (AJAY) Phase II	Till 31.12.2020	Installation of 3,04,500 units of solar street lights in UP, Bihar, Jharkhand, Odisha, and Assam

Solar Study Lamp Scheme for School Going Children	30.09.2019	Lamps provided to student at Rs 100/- in constituencies which have more than 50% un-electrified households.
Wind		
Wind Data Sharing Policy (WDSP)	22.10.2016	NIWE has commissioned 50m, 80m, 100m and 120m height masts with multiple level sensors to collect dedicated wind resource data including wind profile along height.
National Wind Solar Hybrid Policy	14.05.2018	Promote large scale integration of hybrid renewable energy sources for effective land utilization and reduce power losses.
National Offshore Wind Energy Policy	06.10.2016	Already installed offshore wind energy plants of 8.7 GW overall capacity in countries around the world such as Germany, Demark, Sweden, United Kingdom and Belgium.

1.5.Introduction to Energy Sustainability and Environment

1.5.1. Energy Use Trend in Developing and Developed Countries

The Organization for Economic Co-operation and Development (OCED) is an international economic organization comprising of 38 countries. These countries have a high human development index (HDI) and are regarded as developed nations. These nations have a 62.2% share of global nominal GDP. Some member nations of OCED are the United States, United Kingdom, Germany, France, Denmark, etc. The best four developing nations with relatively higher HDI are Brazil, Argentina, China, and India. A comparison is presented in the amount of metric tonnes of oil consumed by developed nations and developing nations.

From the trend, it can be seen that, developed nations on an average have consumed around 6300 Mtoe equivalent of oil for energy production in the last 30 years whereas, developing nations have 2793 Mtoe. This pattern in recent years, especially last 3-4 years have seen developing nations catching up the developed nations in the consumption of energy. China and

India of the developing nations are first and third around the globe in recent years energy consumption data.

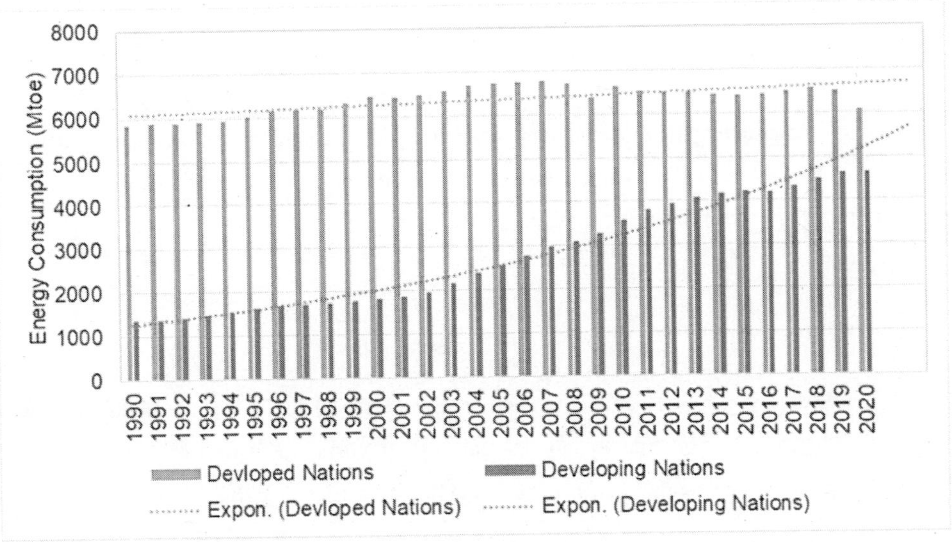

Fig. 1. 14. Energy consumption comparison of developing and developed nations

For the latest data of energy consumption, please scan the QR code

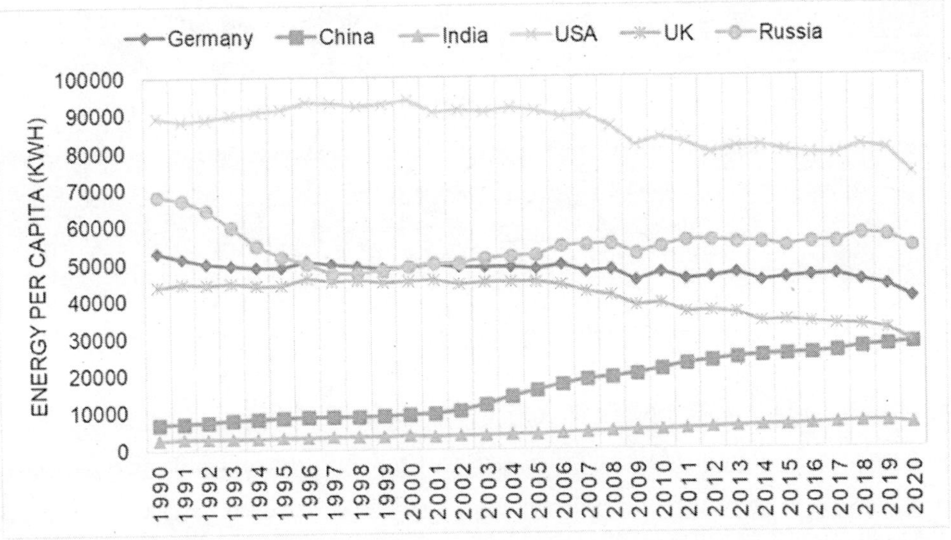

Fig. 1. 15. Per capita energy consumption of different countries

For the latest data of energy per capita, please scan the QR code

1.5.2. Per Capita Energy Consumption

Energy per capita [Wh] = Total population energy consumption [kW·h/yr] × 1,000 /population

The United States has the highest energy consumption per capita, followed by Russia. The major developed nations have brought down their per capita energy consumption by formulating eco-friendly policies and signing accords. Whereas, China is the country with an increase of almost 300% in per capita energy consumption since the start of the millennium. In 2020, India's per capita energy consumption was 6437 kWh as against the world average at 19836 kWh.

1.5.3. Need for Sustainable Energy

One of the most important tasks of the twenty-first century is to tackle the energy issue. Perfect solutions will be difficult to come by, not only because of worldwide variances in governmental and mass support for a sustainable form of energy, but also because of the enormous knowledge necessary to meet the various issues that the global energy environment presents.

Sustainable energy is defined as the energy that meets the energy requirements of the present generation without compromising the energy demands and climate for future generations. The most popular sources of sustainable energy, including wind, solar and hydropower, are also renewable. Due to the widespread availability of oil, petroleum and coal some government officials and public sentiment say these are sustainable kinds of energy, but this is a contentious and debated position. However, most professionals in the energy business believe that even some non-renewable energy sources may be sustainable if utilized in moderation.

1.5.4. Difference Between Sustainable and Renewable Energy

Even between industry specialists and veterans, the terms "renewable energy" and "sustainable energy" are frequently confused. Most sustainable energy sources are also renewable, so there is some common ground between the two. These two concepts, on the other hand, are not synonymous. RE is not always sustainable, and vice versa. Energy must be effectively acquired and supplied in order to be considered sustainable. However, even RE sources, such as biomass, aren't always environmentally friendly. Biomass is renewable because it is biologically generated material derived from natural sources that may be burnt for heat or fuel. However, while biomass is self-renewing, using biofuel produces greenhouse gases, which can harm the ecosystem and, as a result, future generations. As a result, unless this type of renewable energy is properly managed to maximize energy output while reducing negative consequences, it will not be sustainable.

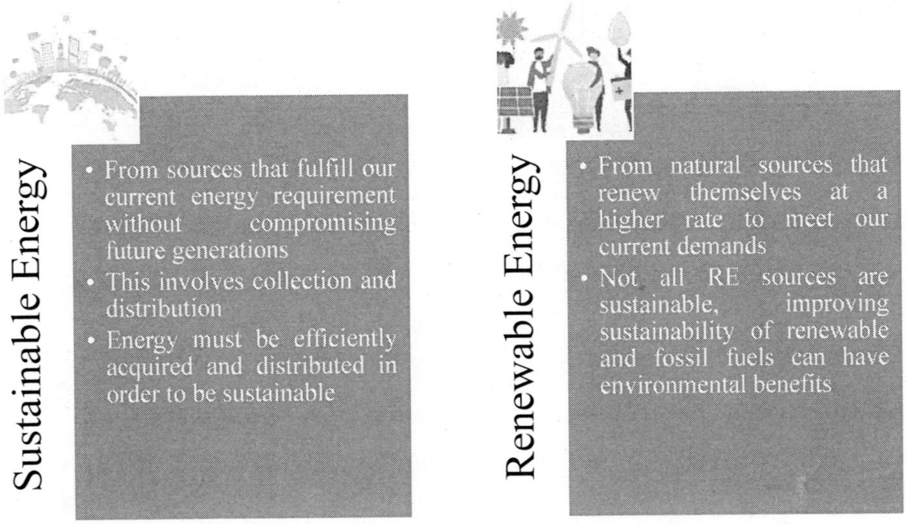

An energy professional one should have an understanding of energy generation, distribution and consumption and the factors that affect it. Such as:

- Where and how can we obtain energy in ways that provide for efficient use, reduce environmental impact and remain cost-effective?
- How is energy distributed and ultimately consumed?
- How are the costs of creation-distribution-consumption measured against the rate of adoption?

1.5.5. Waste to Energy

Organic trash, e-waste, hazardous garbage, inert waste, and other forms of waste are formed as a result of our everyday or industrial operations. Organic waste is defined as trash that degrades or is broken down with time by microbes. Organic wastes are all carbon-based substances, even though they are diverse in composition and degrade at various rates. Organic waste accounts for a major fraction of total waste creation in the industrial, urban, and agricultural sectors, and hence may be utilized to generate energy. The organic waste can be further classified into nonbiodegradable and biodegradable organic waste.

Following are the waste to energy technologies to recover energy from waste [14]:

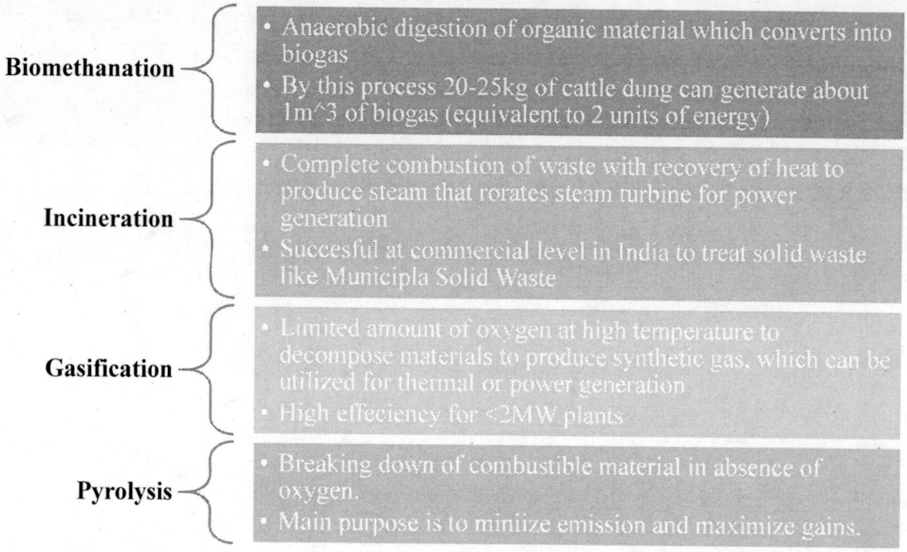

Fig. 1. 16. Various waste to energy technologies

1.5.6. World Energy Policy to Protect Environment

With encouraging signs that energy is becoming more sustainable and broadly available, the world is making progress towards affordable and clean energy (i.e., Goal 7 of sustainable development goals). In poorer nations, access to energy has risen exponentially, energy production and efficiency is improving, and making significant progress in power sector. Nonetheless, improving access to clean and safe cooking fuels and technology for 7.8 billion people, expanding the use of renewable energy outside the power sector, and increasing electrification in remote locations require more targeted effort. The collection of energy and climate-related policies and measures that feed into our modeling is constantly updated and expanded. Internationally power sector policies are being formulated to achieve sustainable development goals. These policies are focused on

- Increased installation of renewable and sustainable energy
- Improve support for carbon capture, utilization and storage
- Stringent pollution emissions limits for industrial facilities above 50 MW_{th} input using solid fuels set at 200 mg/m^3 for SO_2 and NO_x, and 30 mg/m^3 for $PM_{2.5}$
- Mandatory energy conservation construction rules, including a mandate for all new buildings to have net zero emissions by 2030

- Policies promoting production and use of alternative fuels and technologies such as hydrogen, biogas, biomethane and CCUS across sectors.

Some of the steps taken by countries related to electricity sector are listed below [15]:

Regions	Initiatives
United States	• 100% carbon-free electricity by 2050 in 20 states • 30 GW offshore wind capacity by 2030 • Extension to renewable tax credit for solar and wind • Nuclear compensated with zero emissions
Canada	• By 2030 reach nearly 90% non-emitting renewables • Stop power generation from coal-fired plants by 2030
European Union	• Stop power generation from coal-fired plants in 16 nation members • Close nuclear power plants in Germany by the end 2022 • Strengthening national energy plans for offshore wind targets
China	• Target 40% of electricity consumption from renewables • Install over 1200 GW solar and wind energy systems by 2030 • 70 GW nuclear generation by 2025
India	• 450 GW renewables capacity installed by 2030 • 60% of total installed capacity from renewables by 2030
Southeast Asia	• 30% of capacity additions from new and renewable energy sources • 18 GW installed wind capacity by 2030 • Phase out installation of coal power plants

SUMMARY

In this unit, the general idea about energy, its history and technologies, and processes related to energy resources have been discussed. The impacts of energy sources on society, environment are also presented. Further, the role of energy in climate change and related issues and how renewable energy resources are helpful in mitigating this important global issue are also discussed. The development of any technology is based on the promotion of the technologies by various stakeholders including the Government, therefore, the initiatives taken by the government for the promotion of renewable energy are also taken care of.

Short and Long Answer Type Questions

1. Discuss the historical concept of energy in science and its measurements.
2. How can energy resources be classified?
3. What are the different forms of energy?
4. Discuss the merits and demerits of various renewable energy sources.
5. Discuss the Government of India's initiatives to promote renewable energy resources.
6. Energy cannot be created or destroyed. Justify
7. All forms of energy are related to motion. Justify.
8. How scientific innovations and inventions have impacted society and the environment?
9. What are greenhouse gases (GHG) emissions? How does it impact the environment and society?
10. What is the present per capita energy consumption of India, world average and developed nations? What information can be extracted by knowing the per capita consumption of a country?
11. What is the basic difference between renewable energy and sustainable energy?
12. What is world energy policy to protect environment?
13. What is waste to energy? Discuss different technologies involved in waste to energy process.
14. Present the trend used of energy in developing and developed countries. How the projected energy of any country be estimated?

REFERENCES AND SUGGESTED READINGS

1. A. Kumar, "Energy Generation Through Vedas," International Journal of Science Technology and Management, 5, pp. 294-304, 2016.
2. C. Smith, "The Science of Energy – a Cultural History of Energy Physics in Victorian Britain," The University of Chicago Press, 1998.
3. A.M. Helmenstine, "How Does Electrical Energy Work?", Thought Co, 2020, URL: www.thoughtco.com/electrical-energy-definition-and-examples-4119325.
4. Britannica, The Editors of Encyclopedia, "energy," Encyclopedia Britannica, 2021, https://www.britannica.com/science/energy. Accessed 28 May 2022.
5. M. Walker, "Chapter 2: Concept of Energy," Centre of Daily Times, 1986 URL: https://personal.ems.psu.edu/~radovic/Chapter2.pdf
6. NCERT Experts, "Chapter 11: Work and Energy," Physics Book, NCERT, 2021

7. M. Kumar, "Social, Economic, and Environmental Impacts of Renewable Energy Resources," Wind Solar Hybrid Renewable Energy System, Intech Open Press, 2019.
8. IRENA, "Renewable capacity statistics 2021," International Renewable Energy Agency, Abu Dhabi, 2021.
9. United Nations, "Paris Agreement," United Nations, Paris, 2016.
10. IBEF, "Renewable Energy Industry in India Report," India Brand Equity Foundation, 2021.
11. Ministry of Renewable Energy, "Development of Solar Parks and Ultra Mega Solar Power Projects," Government of India, New Delhi, 2017.
12. Ministry of Renewable Energy, "Scheme for setting up of over 5000 MW grid connected SPV power projects under IV of JNNSM PHASE II," Government of India, New Delhi, 2019.
13. Ministry of Renewable Energy, "Grid Connected Solar Rooftop Programme," Government of India, New Delhi, 2019.
14. M. J. Rogoff, F. Screve, "Waste-To-Energy: Technologies and Project Implementation," Elsevier, Amsterdam, Netherlands, 2019.
15. United Nations, "Global Environment Outlook 6 (GEO6) – 2019," New York, USA, 2019.

****

2 Energy Sources

UNIT SPECIFICS

In this unit, the topics related to fossil fuels, various renewable energy resources, and storage systems have been discussed.

- Different types of fossil fuels as energy sources
- Renewable energy resources
- Different energy storage systems

Here, in this unit, the main focus is to describe the various fossil fuels-based energy resources, different renewable energy resources along with storage systems.

RATIONALE

This unit mainly focuses on energy sources for power generation. These energy sources include fossil fuels and renewable energy resources. Different forms of storage systems are also described here.

PRE-REQUISITES

Basic Knowledge of Energy and Environmental Science

UNIT OUTCOMES

List of outcomes of this unit is as follows

U2-O1: An overview of energy sources and energy systems.

U2-O2: Alternative to fossil fuels for environment-friendly power generation.

U2-O3: Sustainable and Environment trade-off of different energy systems.

U2-O4: Energy storage devices

Unit outcomes	EXPECTED MAPPING WITH COURSE OUTCOMES (1- Weak Correlation; 2- Medium correlation; 3- Strong Correlation)					
	CO-1	CO-2	CO-3	CO-4	CO-5	CO-6
U2-O1	3	3	1	-	-	-
U2-O2	2	2	3	-	-	-
U2-O3	2	1	2	-	-	1
U2-O4	1	3	3	-	-	1

2.1. Overview of Energy Sources

Energy is defined as the ability of the system to modify the state or condition of the other systems. Energy exchange is the driving force behind all changes in nature. As a result, energy constantly flows between systems and cannot be generated or destroyed.

An energy system is a system in which energy flows to execute certain specific tasks. In such a system, the energy is converted from one form to another (i.e., the first law of thermodynamics), and some portion is always lost in this conversion process (i.e., the second law of thermodynamics). This measure of conversion efficiency is called as system efficiency. The ratio between energy output and energy input is defined as system efficiency.

2.2. Sources, Transformations, Efficiency, and Storage

Energy is described as the ability to perform work in physics. The varied sources of energy are such as petroleum, coal, sun, wind, and water. These sources, as previously mentioned, can be further categorized as renewable and non-renewable sources. For the energy to be utilized in the useable form, the energy from specific sources would be converted into other forms.

$$\eta = \frac{\text{Energy Output } (e_{out})}{\text{Energy Input } (e_{in})} \times 100$$

$$\eta = \frac{\text{Energy Input } (e_{in}) - \text{Energy loss } (e_{loss})}{\text{Energy Input } (e_{in})} \times 100$$

This conversion from one form to another useable form is called energy efficiency. This is shown in Fig. 2.1.

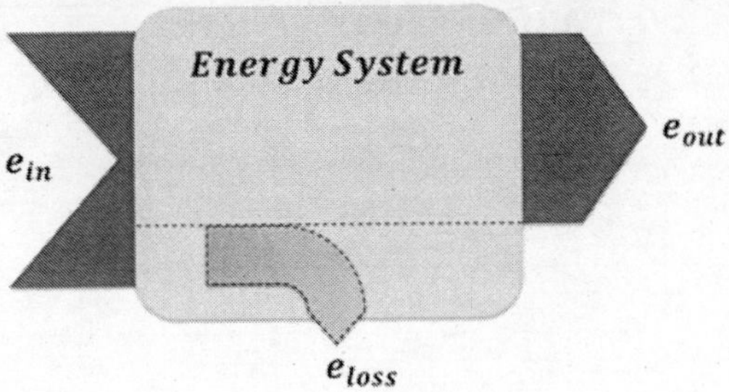

Fig. 2. 1. Depiction of energy and its efficiency

The heat value of an energy source is the amount of heat released during its combustion. Also referred to as energy or calorific value, heat value is a measure of a fuel's energy density and is expressed in energy (joules) per specified amount. The conversion of input energy to output energy flows from primary energy supply to final energy use, and eventually, useful energy flows and provides services. This flow is shown in Fig. 2.2.

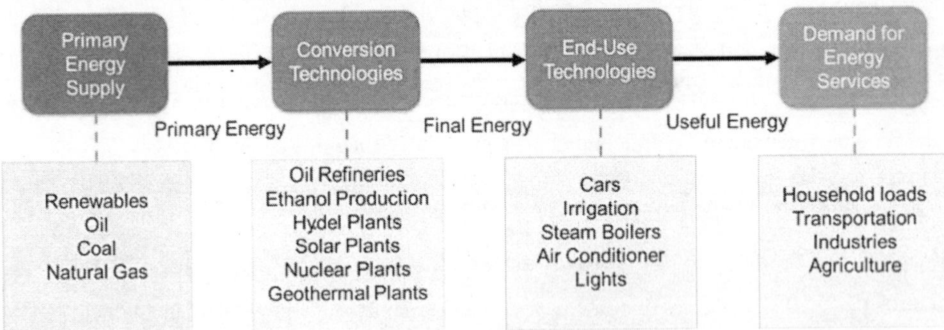

Fig. 2. 2. Conversion of energy from the different sources into a useable form

2.3. Fossil Fuels

Oil, coal, and natural gas are examples of non-renewable fossil energy sources that originated when prehistoric flora and fauna perished and were progressively buried by layers of rock. Different forms of fossil fuels originated over millions of years, based on the combination of organic matter present, how long it was buried, and the temperature and pressure conditions that prevailed at the time.

Today, the fossil fuel industry drills or mines for these energy sources, then burns or refines them for use as heating or transportation fuel. The combustion of fossil fuels accounted for roughly three-quarters of all human-caused emissions during the last 20 years. According to the data available on 31st March 2022, in India total electricity generation capacity from fossil fuels contribute to 59.10% [1]. Fossil fuels are the energy sources formed after millions of year of decay and decomposition. Unfortunately, they account for major carbon emissions and the need of the hour is to reduce the dependence on these resources. In the present scenario, scientists and engineers have a major role in devising alternatives for a cleaner and healthier environment.

2.3.1. Coal

Coal is a flammable material that is generally black or brown in appearance. Coal is mostly made up of carbon-based elements obtained from ancient plants, with some inorganic additives thrown in for good measure. Coal's fundamental chemical element is carbon. Coal has a carbon content of more than 70% by weight in most cases. Coal has a calorific value of between 25 and 35 MJ/kg. Lignite, bituminous coal, and anthracite are the three basic forms of coal. The best grade coal is anthracite, which has a carbon content of more than 90% and the highest calorific value. Bituminous coal is one of the hardest varieties of coal in terms of physical hardness, with a carbon concentration of 70–75 percent. Coal also contains sulphur and nitrogen. Coal accounts for over 37% of the world's electricity supply. Besides electricity, another sector heavily dependent on coal is steel production. An exorbitant amount of 70% of steel production depends on coal feedstock. The downside of burning coal is that it produces 14 billion tonnes of carbon dioxide. Thus, the development of 'clean coal' technologies appears as an immediate solution to this problem.

Coal-fired power plants are still the greatest source of electricity generation and energy-related CO_2 emissions, posing a significant challenge for governments attempting to achieve net-zero emissions while preserving energy security and affordability. Coal accounts for nearly a third of worldwide energy generation, and it will continue to play an important role in sectors like iron and steel until alternative technologies become available. Coal is the greatest single source of CO_2 emissions and the largest source of power generation, posing a particular challenge in the transition to low-carbon energy systems. According to the Internation Electricity Agency's report on (IEA) Net Zero Emissions by 2050 Scenario, all unabated coal production will cease by 2040. Despite the fact that 20 nations have set deadlines for phasing out coal for electricity production, coal is expected to remain an important part of many countries' energy mix. Governments and the coal industry must develop and implement less polluting and more efficient technologies, such as Carbon Capture, Utilization, and Storage, in order for coal to have a future as a cleaner energy source in the decades ahead. Since the Paris Agreement was signed, 21 nations have committed to phase out coal-fired power production from their energy grids, many by 2030. In 2020, these 21 nations accounted for 3.2 percent of worldwide power output and 1% of total CO_2 emissions [3]. The geographical breakdown of worldwide coal consumption reveals that coal's future will be primarily determined in big Asian economies, where energy demand is still expanding in many cases, and coal phase-out objectives have yet to be established.

2.3.2. Oil

Oil here is normally referred to as crude oil and petroleum. These are considered fossil fuels because they are hydrocarbon mixes generated from the leftovers of animals and plants (diatoms) that have lived in a marine ecosystem since the prehistoric era. The remnants of these creatures and plants were covered by layers of sand, silt, and rock over millions of years. The residues were transformed into crude oil or petroleum as a result of the heat and pressure exerted by these strata. Petroleum is short for "rock oil" or "earth oil. Post crude oil extraction from the ground, transportation to the refinery takes place. Thereafter, it is processed into usable petroleum products. Gasoline, distillates such as diesel fuel and heating oil, jet fuel, petrochemical feedstocks, waxes, lubricating oils, and asphalt are examples of petroleum products. More information may be found in Refining crude oil—inputs and outputs.

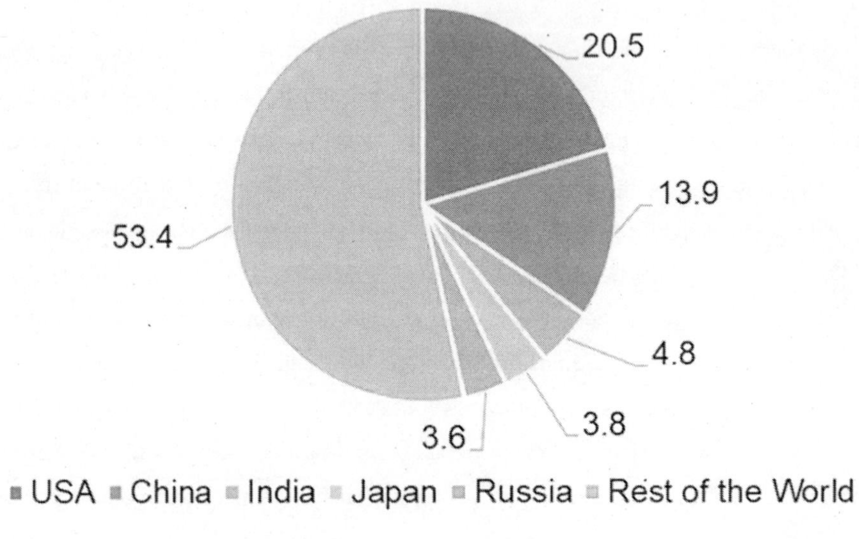

Fig. 2. 3. Petroleum-consuming countries in 2018

Approximately 40% of methane emissions are now attributed to oil production, with the remaining 60% attributed to leaks across the natural gas value chain. More than three-quarters of total emissions come from upstream oil and gas operations, with the downstream segment accounting for the rest. Methane emissions declined by 5% in 2020, owing mostly to lower energy production, however emissions are expected to increase in 2021. Methane emissions fall quickly during the following 10 years under the Net Zero Emissions by 2050 Scenario, decreasing by 75% in 2030 compared to 2020 [4]. This is due to the quick implementation of reducing emissions technology and methods that resulted in the complete removal of all scientifically preventable methane emissions within this time frame.

In India, total petroleum product consumption in March 2022 stood at 19.41 million tonnes, the highest since March 2019, data from the Petroleum Planning and Analysis Cell of the oil ministry showed. Diesel, the most widely used fuel in the country, saw demand rise by 6.7 percent to 7.7 million tonnes, accounting for about 40% of all petroleum product use. Petrol sales increased by 6.1 percent to 2.91 million tonnes, surpassing pre-Covid levels a few months ago.

2.3.3. Oil Bearing Shale and Sand

Shale oil differs from conventional crude in many respects. The former is also called 'tight oil'. It is found in smaller batches and deeper than conventional crude deposits. Its extraction requires the creation of fractures in oil and gas-rich shale to release hydrocarbons through a process called hydraulic fracking. An oil shale is a sedimentary rock that contains kerogen, a petroleum-like liquid that is produced when the rock is heated. Tar sands are made up of clay, sand, water, and a heavy hydrocarbon called bitumen. Tar sands' bitumen may be converted to synthetic crude oil just like kerogen in oil shale. Oil shale and tar sands are referred to as uncommon fuel sources since generating fossil fuels from these resources are unusual and more complex than standard energy production, such as coal mining or oil and gas drilling. There are two ways of extracting shale oil, neither of which has been shown to be feasible. These methods of extraction are underground, open-pit, or strip mining, and Situ method. In terms of global production of shale oil, Russia and the Us ranks as the largest producers in the world. The large-scale production of shale oil in the US turned the country from an importer to a net exporter of shale oil in 2019. Three factors have come together in recent years to make shale gas production economically viable [5]:

- Technological advances in horizontal drilling

- Hydraulic fracturing
- Increase in natural gas prices in the global market.

Oil extraction from shale or tar sands is dirtier than coal extraction. This is especially true in the case of oil shale development, because extracting a barrel of oil from stone requires a tremendous amount of energy. In reality, each gallon of shale oil produced emits 50% more CO_2 into the environment than a gallon of conventional oil. This really is untenable at a time when we need to drastically reduce, not double, our CO_2 emissions. Fuel economy, public transportation, improved urban planning, and a new generation of automobiles are all preferable investments for reducing foreign imports over the next 30 years.

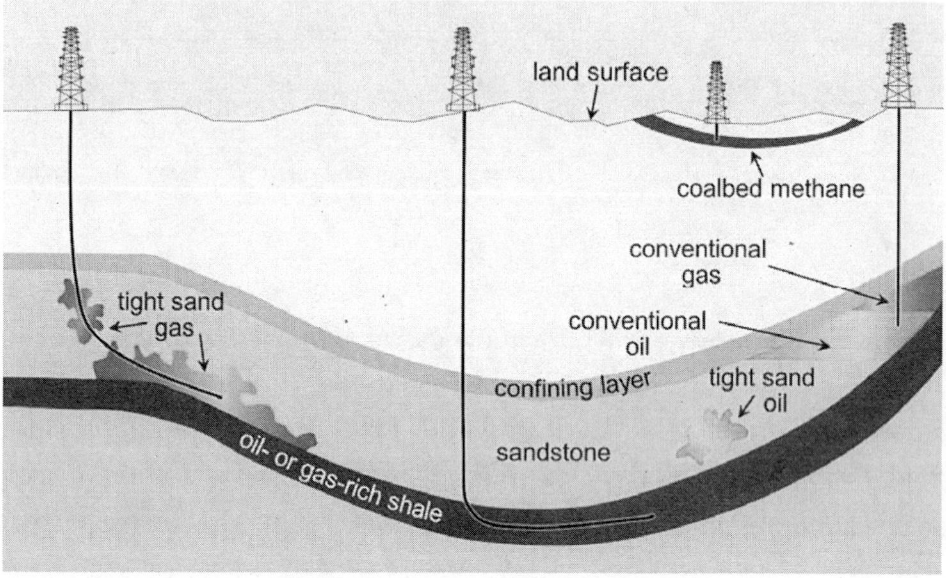

Fig. 2. 4. Conceptual illustrations of types of oil and gas wells

(Source: https://en.wikipedia.org/wiki/Tight_oil)

2.3.4. Coal Gasification

Over the years, many gasification R&D facilities have been developed. Each facility attempts to exploit the chemistry, kinetics, and thermodynamics of coal gasification. Some of these processes have reached a commercial level. A common feature of all processes is that coal is contacted with gasifying agents, mainly oxygen and steam, in a reactor at high temperatures, mostly under a pressurized atmosphere. The solid coal loses moisture, volatile matter, and residual char is gasified, leaving ash as the residue. Carbon dioxide and heat are produced in-situ by the combustion reaction and further generated heat utilized to drive the endothermic gasification reactions. The art of gasification lies in balancing the exothermic and endothermic reactions while maintaining the required reactor temperature. Although the three main types of gasifiers (i.e., entrained flow, fluidized bed, and moving bed) can be used to gasify coal, gasifier efficiency and stability are ensured under a range of values of certain characteristics of the coal.

The coal gasification technology has attracted increasing importance internationally due to the low production cost of hydrogen and sustainability. In the United States, 95% of hydrogen is produced by a reaction between a methane source, such as natural gas, and high-temperature steam (700°C–1,100°C), referred to as steam methane reforming (SMR). About 4% is produced through coal gasification, and 1% is produced from electrolysis. China, the biggest producer of coal on the planet, also shares the highest production of hydrogen (20 million tonnes per year), mostly from coal gasification. Currently, 70% of China's hydrogen comes from nearly 100 coal gasification plants installed in the country. The overall production cost of hydrogen from coal in China is also lower ($0.95-1.90/kg H2) compared with the natural gas-derived hydrogen ($1.27-2.37/kg H_2) and green hydrogen ($3.94-5.54/kg H2) [6]. Despite the low production cost, the emission of CO_2 from the coal gasifier is huge. It is estimated that 8 kg of coal produces about 1 kg of hydrogen in China and emits 20 kg of CO_2 in the environment.

The Paris Accord in 2016 gave governments around the world, an opportunity to make and implement commitments regarding reducing greenhouse gas emissions. Several nations in this attempt made the ambitious commitment of achieving net-zero greenhouse gas emissions by 2040-2060, the other countries planned to reduce the emissions in a phase wise manner. These commitments have been a part of clean production and decarbonization strategy. Besides public sector, many private companies vouched to net zero goals between 2030-2050. To achieve these targets, low-carbon looks like a viable solution for zero-emission energy services.

Energy cannot be generated or destroyed, according to the rule of conservation of energy. It can, however, be changed from one form to another. The overall energy of an isolated system remains constant when all kinds of energy are considered. Moreover, it can be further elaborated that energy remains constant for a closed system, i.e., isolated from its surroundings when all the energy sources are considered.

2.4. Past, Present, and Future

Fire has been utilized by humans for thousands of years, dating back to the late Stone Age. Cave inhabitants kept their caverns warm with fires that were kept blazing continuously, according to archaeologists. Fire was later employed for more complex purposes, including as cooking, lighting, heating pottery, smelting metal, and creating glass. People used the energy stored in their muscles to undertake activities such as hunting, gathering food, and constructing a shelter in ancient times. To aid in these endeavors, simple tools were created. Wood was used to make the first tools. They were afterward fashioned of metal. Human force was also employed to construct the first boats, which were subsequently propelled by poles and oars. Because some of the tasks were too tough to do with human strength alone, humans turned to animals like oxen. Animals provided the energy needed to execute a variety of jobs, including transportation, assisting in the carrying of large objects, plowing lands for agriculture, and pumping water for irrigation. Plows have been pulled by animals for thousands of years; the first plows date back to 4000 BCE in Mesopotamia [7].

It has been documented that the Chinese were using coal as early as 1000 BC. to bake porcelain. Coal had gained prominence as a heating fuel by the 1600s. It was also utilized in breweries, glass factories, brick factories, and a variety of other industries. Coal's popularity continues to rise. It was first extracted from shallow near-surface deposits. As the coal was used, miners proceeded to construct mine shafts in order to reach deeper into the earth, where there were numerous coal resources. This brought about the Industrial Revolution, and in 18th century as coal was utilized in tall furnaces.

The electromechanical devices were built to aid human civilization. To harness wind energy along the coastlines, sailboats were invented for exports and trading purposes. In the seventh century, windmill was designed to utilize wind power to do work on land. In the 20th century, wind-powered grain mills and sawmills were phased out in favor of more efficient gear. Windmills are still used by some farmers nowadays to draw water, clear floodwater, and pull up groundwater for crop irrigation. Wind turbine generators, or high-powered windmills, are being employed to generate energy. The first electric light was created by Sir Humphrey Davy, an Englishman, in 1802. He operated with a battery and an electric spark between two conductive materials. At the end of each conductor, he put a little charcoal rod. His device was then placed inside a glass globe. Later on, Thomas Edison invented the modern bulb using a very fine carbon wire to control the bulb's glow. Inventors utilized this technology to illuminate entire cities over the following few decades. In 1879 electric motors were first used as power trains for energy production.

Energy has shaped human civilisation, and the usage of energy sources, particularly fossil fuels, has had a significant impact on people's living situations. The range of human applications for diverse energy sources has expanded to the point that modern life would be difficult to imagine without them. To put it another way, human existence has become increasingly reliant on energy consumption, yet the benefits it provides come at a high cost to the environment and vulnerable groups. Energy is a driving force that promotes a range of desired results, such as personal comfort and well-being, social welfare, economic growth and prosperity, and technological, industrial, and sustainable development [8]. Energy technology provides countless opportunities for scientific research and innovation, particularly when applied to renewable and alternative energy sources. Trends to date indicate a positive step in developing a more energy-efficient tomorrow. Some of the future trends for energy efficiency and sustainable development are as shown in Fig. 2.5.

Fig. 2. 5. Future trends for energy efficiency with different sources

2.5.Remedies and Alternatives for Fossil Fuels

World energy demand is projected to increase exponentially by the year 2050 [9]. Continuous resource depletion and rising greenhouse gases emission will necessitate the shift from fossil fuels to renewable sources for energy consumption [10]. In order to ensure human welfare, the world is advocating the use of renewable energy. The concept of sustainable development will become a reality when human welfare and growth in the face of high energy consumption are balanced with an environment devoid of carbon emissions. Renewable energy technologies present the advantage of clean and abundant energy from self-renewing sources such as the sun, wind, earth, and water. Renewable energy sources have a twin-fold advantage of completing the world's energy demand and providing energy security. The COVID-19 pandemic posed not only a health crisis but had a major implication on global economies, energy usage, and CO_2 emissions. In the first quarter of 2020, when most countries were under severe lockdown, global coal demand was the hardest hit, and oil demand was hit strongly by 5% in the first quarter of 2020 [11]. However, IEA reports that the renewable energy sources posted a positive growth demand of 1.5% in quarter 1 of 2020, driven by large installed capacity and priority dispatch. Among the various regions in the world, Asia shows promising results in terms of installed capacity of renewable energy, as presented in Fig. 2.6. In recent years, India has paved a path for sustainable energy sources. India aims to attain 175 GW of renewable energy, which includes 100 GW from solar energy, 10 GW from bio-power, 60 GW from wind power, and 5 GW from hydropower plants by the year 2022.

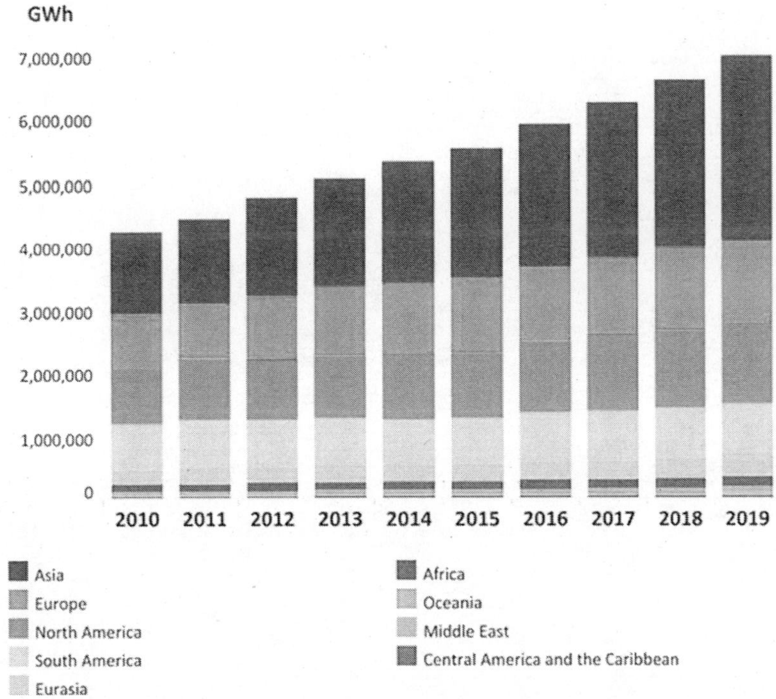

Fig. 2. 6. Trends in renewable energy by region

For more information regarding renewable energy trends in different regions, scan this

The subsequent section discusses details of various alternatives to fossil fuels such as solar energy, wind energy, hydro energy, tidal energy, geothermal energy, nuclear energy, and biomass energy.

2.5.1. Biomass

Biomass comprises organic matter originating from plants, trees, and crops, and is primarily the collection and storage of the sun's energy via photosynthesis. The energy produced from biomass or bioenergy comes directly from the crops, or from residues generated through the processing of crops for food or other products. Though biomass can be directly utilized to produce energy, it can also act as a feedstock to be converted various liquid or gas fuels, also known as biofuels. The recent strategy of governments to develop biorefinery to convert biomass into clean energy biofuels is gaining popularity. The inter-conversion process includes processes such as combustion, pyrolysis, gasification, and anaerobic digestion. The utilization of biomass-related fuels has the capacity to fulfil energy demands and foster socio-economic developments for several nations. In contrast to the benefits of biomass energy, such as reduced dependency on fossil fuels, lower levels of greenhouse gas emissions, and reduced smog, there are certain limitations to biomass-to-energy facilities [12]. Biomass fuels have low energy densities and can emerge as a cost-prohibitive source.

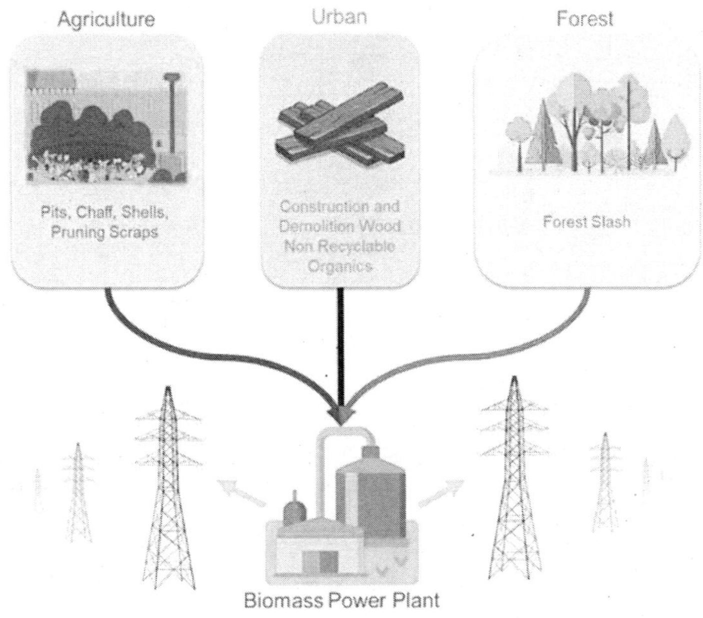

Fig. 2. 7. Illustration of biomass energy

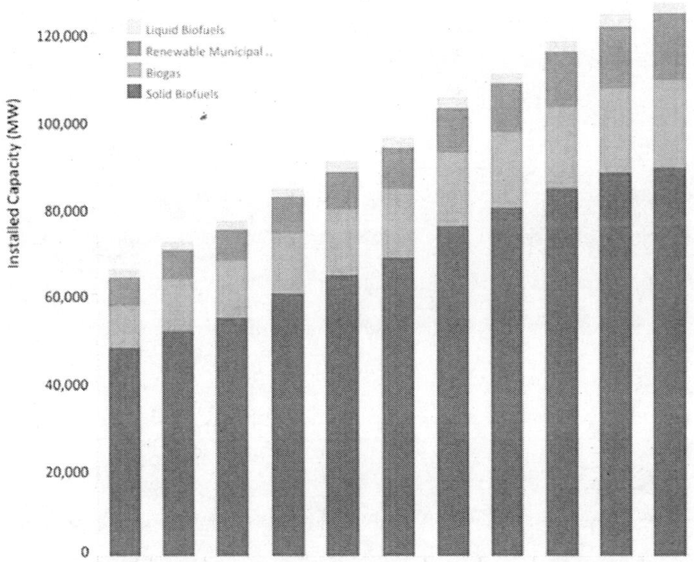

Fig. 2. 8. Installed capacity transformed from heating or power generation from biomass from 2010-2020

For more information on the installed capacity of biomass plants, scan this

Bioenergy currently accounts for around 10% of the global energy supply. Research suggests that bioenergy has the potential to offer from 10% to more than 60% of the global energy supply. It is estimated that by the year 2060, the production of bioenergy will expand from the current level of 56 EJ to 145 EJ [11]. There are several sources of bioenergy, such as crops, algae, and lignocellulosic biomass. Wood crops, sugar sources, and herbal plants are other options considered for the production of bioenergy. The installed capacity of bioenergy is used either for heating or for power generation from biomass- liquid biofuels, solid biofuels, biogas, and renewable municipal waste.

2.5.2. Wind Energy

Wind energy is another highly useful source of energy obtained through the conversion of wind energy by wind turbines into a useful form. Windmills are used for mechanical power, and wind pumps are used for water pumping or drainage. The earliest use of wind turbines for generating electricity was found at the beginning of the 20th century. Since then, the technology has drastically improved, and in the 1990s, wind energy remerged as an important renewable source. The process of generating electricity is such that the wind requires the kinetic energy of the moving air, which is converted to mechanical energy and then to electrical energy. The design of the wind turbine should be such that it maximizes the energy capture over a range of wind speeds and minimizes the costs of wind energy, taking all parameters into consideration. The wind turbines are onshore and off-shore technology. The onshore wind turbines are at times grouped together into wind power plants, also called wind farms. The off-shore technology is less common and less mature than the onshore one; it also incurs a higher investment. However, the motivation for developing off-shore wind technology largely includes- higher quality wind resources at sea and the ability to use larger wind plants and build larger power plants.

In the early 80s, the United States excelled in establishing wind farms. The operation of setting turbines took at a large scale, each generating energy between 80 to 200 kW. Among the European nations, Denmark became the pioneer of wind energy. At the end of the twenty-first century, Germany took the lead in producing wind energy. In the last few decades, governments of several nations have invested heavily in generating energy through wind primarily because it is a clean energy source and limits the need for fossil fuels. By the end of 2016, the world capacity of wind energy was 487 GW. This was around 1.9% of the installed capacity of power stations, including water, coal, natural gas, and nuclear. The world wind power production is presented in Table 2.1. The global future of wind generation looks bright. IRENA estimates global wind generation to be around 732 GW by 2020. In terms of onshore generation, IRENA expects 100 GW more generation by 2021. However, building a profitable wind generation depends on several factors, such as local geography, which decides wind capacity.

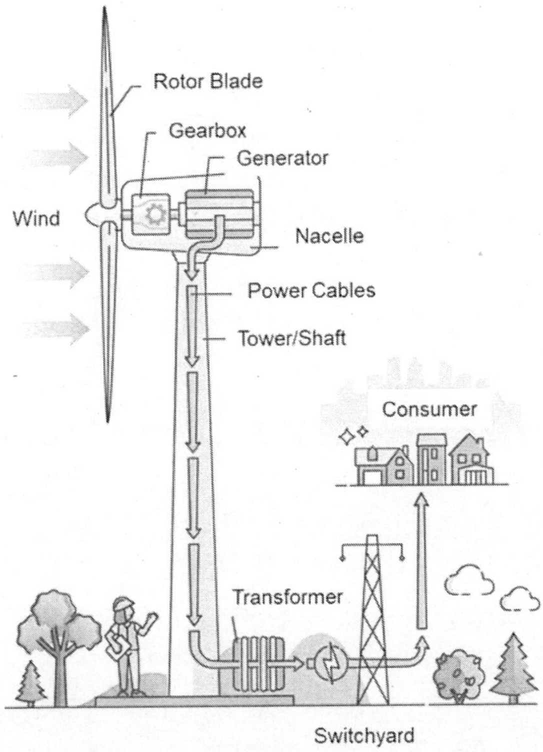

Fig. 2. 9. Components of a wind energy system

Table 2.1. World wind power production

Region/Country	Installed rated power in GW
China	342
USA	139
Germany	64
India	42

Spain	29
UK	26
Brazil	19
France	18

2.5.3. Solar Energy

The source of solar power is the sun, this energy is converted into thermal or electrical energy. Solar energy is one of the cleanest and most abundant renewable energy sources available. Solar technologies can harness this energy for a variety of uses, including generating electricity, providing light or a comfortable interior environment, and heating water for domestic, commercial, or industrial use. The most popular renewable energy source across the world is solar energy; it uses the sun's energy to produce electricity via solar photovoltaic systems. The technology for solar energy has undergone momentous evolution, and this is evident in numerous systems installed over the last decade. Solar energy can be harnessed using the two most commonly used technologies such as solar thermal plants and solar photovoltaic systems.

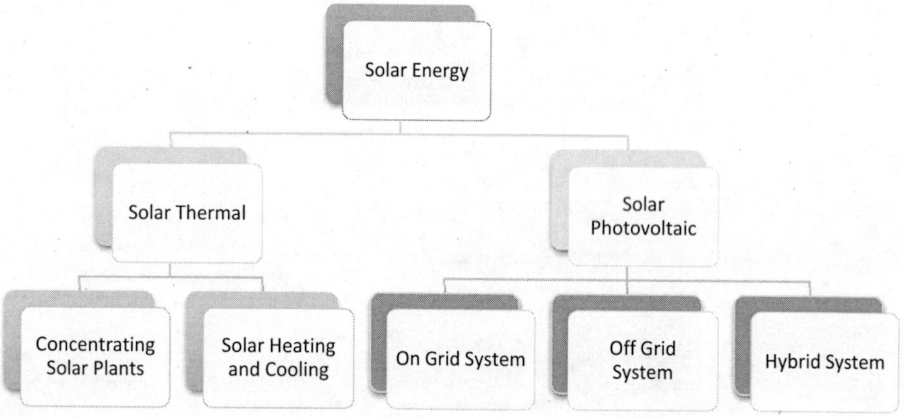

Fig. 2. 10. Classification of different forms of solar energy

Concentrating solar power (CSP) plants use mirrors to concentrate the sun's energy to drive traditional steam turbines or engines that create electricity. The thermal energy concentrated in a CSP plant can be stored and used to produce electricity when needed, day or night. Some CSP configurations are parabolic trough, compact linear Fresnel reflector, power tower, and dish engine. Solar heating and cooling (SHC) technologies collect the thermal energy from the sun and use this heat to provide hot water, space heating, cooling, and pool heating for residential, commercial, and industrial applications. These technologies displace the need to use electricity or natural gas.

The solar photovoltaic (PV) systems perform the function of directly converting solar energy into electricity. For this purpose, the building block of a PV system is the PV cell. Together the PV cells make the PV module, ranging from 50 to 450 W. PV module, along with other components such as inverters, batteries, and mounting system, form a PV system. The PV systems are used to provide power as low as a few watts to as high as tens of megawatts. The most popular solar PV systems are silicon-based. The non-silicon semiconductor materials and film module are gaining importance. The PV systems come with many advantages. First, it is a very modular technology, it can use direct sunlight and can also diffuse components of sunlight, i.e., the PV system can produce power even when the sky is not clear. The PV systems are classified into off-grid, grid-connected and hybrid PV systems as shown in Fig. 2.11. The off-grid system has proven beneficial in unelectrified areas of developing economies, they are widely being used for village electrification. The grid-tied PV systems use an inverter for converting electricity from direct current to alternating current, thereafter supplying the electricity to the grid.

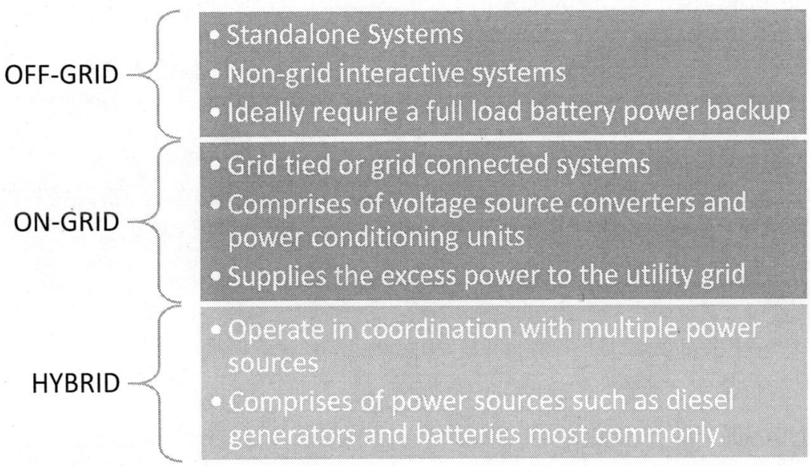

Fig. 2. 11. Different types of solar photovoltaic systems

2.5.4. Nuclear Energy

Nuclear energy is the energy in the nucleus, or core, of an atom. Nuclear energy can be used to create electricity, but it must first be released from the atom. "Thermal" power plants convert heat into electricity using steam. Nuclear energy is viewed as an extremely important component of energy security and economic development. Considering the current world energy needs, it is an important pillar, with 6 GW of additional nuclear capacity connected to the grid in 2020. New projects of around 4.8 GW were launched to ensure long-term operations in the countries. The advantages of providing high power supply capacity and low fuel levels requirement make it a popular source for meeting world energy demands. Along the lines of renewable energy sources, nuclear energy too can be considered a major source for the decarbonization of economies because it is based on low carbon technology. Research also highlights a significant correlation between the use of nuclear energy and managing climate change [13]. Despite the benefits of actually reducing global warming, nuclear energy comes with its downsides, such as radioactive waste generation and safety concerns. However, with nuclear waste management regularly controlled and monitored by the International Atomic Energy Agency, the risks are somewhat under control in the current scenario [14]. At nuclear plants, nuclear fission takes place which involves the splitting of atoms, and subsequent release of heat. Repetition of this process results in a chain reaction. For fission to happen, uranium is used as a material. The generated boils water and creates steam, the steam turns the turbines. The spinning of turbines turns the generator and the magnetic field produces electricity.

Nuclear plants have a complex working mechanism. The plant consists of a containment structure that holds the reactor. The plants have a deep pool for containing nuclear fuel when not in use. Nuclear plants are designed to produce electricity from steam. This steam is utilized in different ways, either with a Pressurized Water Reactor (PWR) or Boiling Water Reactor (BWR). The splitting of atoms also results in radioactivity. Radioactivity management is an essential part of nuclear plant safety management.

2.5.5. Hydropower energy

Hydropower is the energy source derived from the energy of moving water. The energy is captured and converted through the use of turbines. The most common form of hydropower is through dams, although harnessing energy from waves and tides is gaining popularity. The flow of water in rivers, driven by gravity to move from higher to lower elevations, is utilized to generate hydropower. Hydropower plants are categorized into three types- Run-of-River (ROR), storage reservoir, and pumped storage plants. A ROR plant mainly draws energy from the flowing river [16]. The general profile of the plant varies according to the local river flow conditions, and significant daily, monthly or seasonal variations may occur. The storage hydropower plant differs from the former as it stores water for later consumption, reducing its dependency on the inflow. Pumped storage plants are not energy sources, rather are storage devices, where water is pumped from a lower reservoir to an upper one, usually during off-peak hours and during the daily peak load period, the flow is reversed to generate electricity. The pumped storage plant has the capability to provide large-scale energy storage system benefits and is the largest capacity form of grid energy storage, widely used by various countries. Hydropower energy is one of the best conversion renewable energy sources. Electricity generation from hydropower has shown promising results. It accounted for around 17% of the total electricity generation globally between 2010 and 2016 [17]. With the rise in electricity consumption, the absolute value of electricity from hydropower also increased.

2.5.6. Tidal and Wave Energy

Tides and waves are two examples of environmental phenomena occurring in the ocean. While they share a connection to bodies of water, their capacity to produce energy differs in a number of ways, including production, power, and dependability.

The term "tides" refers to the rise and fall in sea level brought on by the moon's and the sun's gravitational pull on the planet. They may happen in other systems whenever a gravitational field is present and are not only restricted to the seas. Additionally, while the majority of the earth is subject to the sun's gravitational pull, this is less obvious in water. Given that it is far closer to the earth than the sun, the moon itself has a more noticeable impact on the tides. A diurnal or semi-diurnal tide, consisting of one or two high and low tides per day, is experienced by shorelines. The two primary forms of tidal energy are 1) kinetic energy, which is derived from the currents of changing tides, and 2) potential energy, which is derived from the variations in height between high and low tides. One benefit of using tides as a source of energy is that they are more dependable since they are dependent on the moon's gravitational pull and can therefore be anticipated. However, the drawback of tidal energy is that it is not a continuous source of energy, with energy provision of 6-12 hours at a time. This intermittent nature of energy reduces its reliability to a large extent. The other disadvantage is that harnessing the energy can disrupt the habitats of marine animals. Often, turbines used for energy generation have been a cause of the loss of aquatic life. Having said this, the ability of tidal energy as a source of electricity is a cleaner option than using fossil fuels.

The definition of wave energy, commonly referred to as ocean energy, is the energy obtained from ocean waves. Waves are regarded to be the force that is traveling over the ocean's surface as a result of the wind, which also causes surges of energy from these waves. Since there are no land masses to withstand the force of the wind, they produce wind waves that are most effective on ocean surfaces. These waves, which can be classified as capillary waves, ripples, seas, or swells, are frequently visible on the ocean's surface but also freely occur on lakes, rivers, and canals. The shape of the waves formed is determined by the changing patterns of the speed and duration of the wind. The motion of the oceanic waves contains kinetic energy, this is captured by the current wave technologies. The wave energy is harnessed using off-shore and on-shore systems. Off-shore systems work in a way that they utilize either pumps or hoes for collecting energy through rotating turbines. Whereas, onshore systems are built along shorelines and harvest energy from the waves. Wave energy comes with the advantage of being replenishable and sustainable. With the use of appropriate technology, the energy harnessed can be made available to the nearby communities. However, the efficiency of wave energy remains in question often. Thus, equipments should be built in a way that they withstand corrosion from saltwater and other damages. Most technologies related to wave energy have found to be cost-effective, however, it is not as cheap when compared with other energy generating systems.

2.5.7. Hydrogen Energy

It is a known fact that the energy produced from carbon-containing sources will lead to environmental pollution. On the alternate front, there are renewable energy sources, in particular solar, wind, and tidal sources. However, the application of these sources is limited by the duration and nature of natural phenomena. Therefore, a source that ensures an uninterruptible energy supply is required. Metal-ion batteries, redox batteries, and hydrogen cells have exhibited promising results. Fuel cells (FC) produce energy through the process of oxidation of hydrogen-containing fuels by oxygen. Several researchers consider it a renewable source because hydrogen utilized here is generated by biomass conversion or alcohol. The benefit of FCs is that they are environmentally friendly and efficient. In order to reduce carbon emission, the use of hydrogen supply to generate electricity, especially for remote and isolated areas. In the current scenario, the global demand for hydrogen is around 115 million tons per annum. This demand is for various purposes, such as ammonia production and oil refining. Among these, the share of final energy consumption is around 3%. It is estimated that the global demand may increase almost six times, to 700 million tons, by 2050. The share of hydrogen for final energy consumption is predicted to reach 24% from the current levels [18]. This rise in demand will primarily be in the transport sector and for electric generation.

2.6. Sustainability and Environmental Trade-Offs of Different Energy Systems

Energy, as we know, is used to provide energy services of various types which include provision of electricity, heating, cooking, transportation and industrial purposes. The complete cycle of energy is complex and includes obtaining energy from various sources which include the renewable and non-renewable sources. Most certainly, energy services allow for a good standard of living and promotes overall development.

The energy sector is witnessing multiple challenges on the economic, political, and technological fronts. These challenges are due to widely increasing demands which can be linked to the different dimensions of a sustainable energy system which include environmental sustainability, security of energy supply, and economical sustainability. The debate between the environment and the economy has been ongoing for a long time. Different stakeholders have different viewpoints. Making the energy sector greener is a viable option for achieving a win-win situation in terms of environment and economy. It is also a priority agenda for governments across the world and has been achieved somewhat by the developed economies. Balancing trade-offs is a key challenge when forming energy policies. The 'common but differentiated responsibilities' principle was coined at the Earth summit in 1992 had been in question for a long time, particularly by the developing economies. Thus, after much deliberation, United Nations Framework Convention on Climate Change passed the 'common but differentiated responsibilities and respective capabilities' that acknowledges the different capabilities and different responsibilities of the countries to deal with climate. Turning towards the greener option in the energy sector is a step in this direction. Despite this fact, several countries are using energy in an unsustainable manner. Apart from the economic and environmental challenges, there is a societal challenge involved in achieving energy sustainability. A few of the challenges include societal inequities, over consumption of resources, climate change and high prices of energy or limited energy affordability. There is a lack of uniformity when it comes to the pricing of energy, the prices are skewed by taxes and incentives. In addition, the standard of living, rising population and level of urbanization affect energy sustainability to a large extent. Considering the ongoing development with limited wealth, education and technology, this challenge is bigger for developing nations.

Thus, a larger emphasis needs to be placed on clean energy fuel considering the trade-offs between environment, economic and social factors. There are several **'co-benefits'** of clean energy systems [19]. Clean energy will eventually limit the impacts of climate change, make countries resilient to deal with man-made or human-induced disasters. A direct benefit would be reducing greenhouse gas emissions. This in turn would help manage the adverse effects of global warming. Well-planned energy efficiency programs can reduce energy use, promote clean energy and cut consumers' energy bills, translating into greater financial resilience to future shocks. However, energy sustainability is not without its own challenges. In order to strike a balance between environment and energy, the primary step is a comprehensive and meaningful assessment of the environmental and ecological impacts. For this the overall life cycle of the energy systems need to be considered, it starts with the harvesting and processing of energy to its utilization to its ultimate disposal. Researchers use the concept of life-cycle assessment to study various systems including energy processes.

Table 2. 2. Description of different sources of energy and corresponding fuel

Source	Description	Fuel
Onshore wind	Wind turbines capture the kinetic energy of the air and convert it into electricity via a turbine and generator.	Wind
Off-shore wind	Off-shore wind turbines operate in the same manner as onshore systems but are moored or stabilized to the ocean floor.	Wind
Solar photovoltaic	Solar photovoltaic cells convert sunlight into electrical energy through the use of semiconductor wafers	Sunlight
Solar thermal	Solar thermal systems use mirrors and other reflective surfaces to concentrate solar radiation, utilizing the resulting high temperatures to produce steam that directly powers a turbine. The three most common generation technologies are parabolic troughs, power towers, and dish-engine systems.	Sunlight
Geothermal	An electrical-grade geothermal system is one that can generate electricity by means of driving a turbine with geothermal fluids heated by the earth's crust. Deep geothermal generators utilize engineered reservoirs that have been created to extract heat from water while it comes into contact with hot rock and returns to the surface through production wells.	Hydrothermal fluids heated by the earth's crust
Biomass	Biomass generators combust biological material to produce electricity, sometimes gasifying it prior to combustion to increase efficiency. Also, Biomass plants generate electricity from landfill gas and anaerobic digestion.	Agricultural residues, wood chips, forest waste, crops, and urban waste
Biofuels	Liquid fuels manufactured from various feedstocks	Corn, sugarcane, vegetable oil, and other cellulosic material
Hydroelectric	Hydroelectric dams impede the flow of water and regulate its flow to generate electricity.	Water
Ocean Power (Tidal and Wave)	Ocean, tidal, wave, and thermal power systems utilize the movement of ocean currents and heat of ocean waters to produce electricity.	Saline water

2.7. Possibilities for Energy Storage and Regeneration

The growing use of renewable energy sources leads to a growing need for energy storage facilities that store energy for later use. More and more sun and wind are becoming key sources of energy. In mountainous regions, the most cost-effective method is to use water reservoirs as batteries, other regions must find a different way. Hence, for effectively utilizing the generated energy using various sources at specific requirement, storage and regeneration is of utmost importance. Some of the systems incorporating energy storage are discussed as follows:

2.7.1. Pumped Storage Hydro Power Projects

In the present scenario, pumped-hydro storage (PHS) and battery storage can be considered as the front-runners PHS. They are often termed as the oldest and most mature large-scale storage technology, accounts for 96% of the installed global energy storage capacity. Data indicates that around 169 GW of pumped storage capacity is installed worldwide. Among the various countries, China is leading with 32.1 GW followed by Japan and USA at 28.5 GW and 24.2 GW, respectively. PHS is an advanced technology based on pumping water to an upstream reservoir during off-peak or the times that there is redundant electricity produced by renewable energy sources (RESs). This stored electricity is released through the hydro turbines when required. The working of PHS plants is such that they use two interconnected reservoirs with one at a higher elevation than the other. In times of surplus energy, water is pumped to the upper reservoir and, in times of excess demand, water from the upper reservoir is released. Thereafter, electricity is generated as the water passes through reversible Francis turbines on its way to the lower reservoir. This whole process goes through a repetition with an overall cycle efficiency of about 80%. In the case of fixed-speed pumped storage plants, power regulation happens while the plant is generating electricity, on the other hand, with the state-of-the-art variable speed technology, power regulation in specific ranges is possible while generating and while pumping, providing additional flexibility to support the grid stability. The technology of PHS is well suited for large-scale applications, allowing to cope with the intermittency of RESs.

Scan this for more information

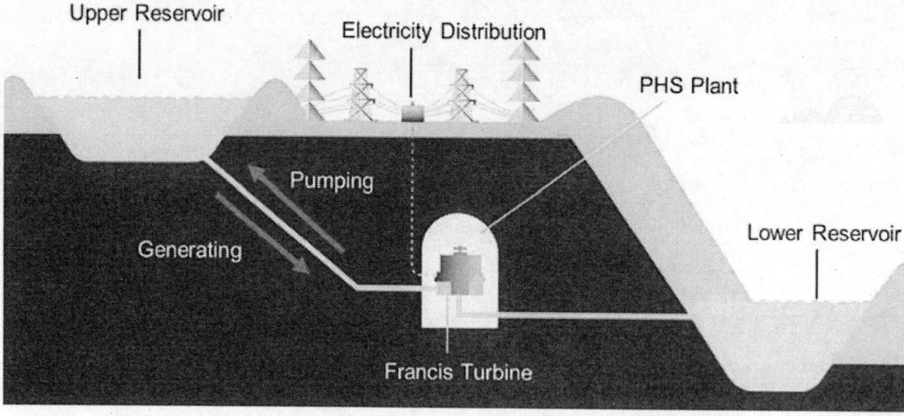

Fig. 2. 12. Depiction of a pumped hydro storage plant

The benefits of PHS are shown in Fig. 2.13.

Large Scale

- This is the attribute that best positions pumped hydro storage which is especially suited for long discharge durations for daily or even weekly energy storage applications

Cost-effectiveness

- Because of its lifetime and scale, pumped hydro storage brings among the lowest cost of storage that currently exist

Reactivity

The current technologies provide response times that are counted in seconds or even milliseconds in the case of variable speed technology

Multifunctional

Water management, irrigation control for agriculture, water distribution and water waste control

Renewable and Sustainable

- pumped hydro storage plants have a lifetime of more than 40 years for the electromechanical equipment and 100 years for the dam. They do not need to be located near an existing river and can therefore be located where needed to support the grid

Fig. 2. 13. Benefits of pumped hydro storage system

2.7.2. Superconductor-Based Energy Storages

Superconducting Magnetic Energy Storage (SMES) is another simple energy storage system which makes use of the dual nature of electromagnetism. The mechanism of SMES is such that the electric current creates a magnetic field followed by changes in the magnetic field which then creates an electric field and voltage drop. The magnetic flux plays an important part as it is a reservoir of energy. Superconducting wires do not deliver energy when conducting a current, so a coil made with those materials maintains the current, and the magnetic flux can be stored. The magnetic flux is a reservoir of electrical energy. The energy is stored/delivered when a controller changes the current, increasing or reducing it, a voltage appears in the terminal, which is regulated by the rate of change of the current, and can be adjusted by the regulator delivering or catching energy to or from the external circuits. Typical power generation ranges from 100 kW to 10 MW with an efficiency of 95%. Some of the main applications of SMES are voltage control and reactive power compensation, improve transient stability of the grid, and uninterruptable power supply (UPS).

SMES offers several advantages when compared to other technologies, namely:

- High round trip efficiency: 90-95%.
- Long lifetime: 30 years
- High power, only limited by electronics and electrical isolation.
- Ready to operate: in a few ms.
- Very robust: can be overloaded as much as the electronics allows.
- Very flexible for hybridization: can be included in an electronic buffer with any other large capacity energy storage system improving their availability speed and their peak power.
- Symbiosis: SMES can take advantage of external resources such as cooling in industrial polygons, hospitals, liquid nitrogen carriers, etc. It allows for reducing the cooling and HTS wire investment cost and enhancing its efficiency.
- Environmentally friendly: specific geometries drastically reduce the stray field bellow any determined level. SMES does not use materials complex for recycling.
- No critical raw materials which are not dispersed and can be recovered after 30 years of service.

2.7.3. High-Efficiency Batteries

Batteries are frequently used as energy storage. Batteries must be kept in a state that allows for longer-term storage since they provide their own set of difficulties. The batteries often drain since they self-discharge and might become inactive after a prolonged time of storage. High humidity is another element that contributes to battery degeneration. There are three main battery kinds that may be used by consumers. They are lithium ion, nickel metal hydride, and alkaline. Each has distinct advantages and disadvantages. Zinc serves as the negative electrode and manganese dioxide serves as the positive electrode in an alkaline battery. The battery discharges, consuming both chemicals. Alkaline batteries are, therefore, disposable, single-use batteries. They cannot be securely recharged after being discharged. On a per-unit basis, alkaline batteries are the least expensive. They deliver dependable performance from beginning to end. In other words, when an alkaline battery is totally depleted, there is a noticeable decrease in power. The recycling of alkaline batteries is a never-ending task.

The cost of NiMH batteries is higher than that of alkaline batteries. Before it runs out, a standard battery may be charged roughly 500 times. The performance of these batteries diminishes with discharge, and with time, there is a noticeable loss of power, which is another drawback. Additionally, NiMH batteries require far longer charging times than conventional batteries.

Li-ion batteries, lithium-ion batteries are the most recent of the three primary types of batteries. These batteries are rechargeable in nature and are used in portable gadgets. Li-ion batteries are most frequently seen in mobile phone batteries. Although Li-ion batteries are the most costly of the three, their main benefit is that they often pay for themselves relatively rapidly. The number of recharges for a decent Li-ion battery is twice as many as for a NiMH battery. The continuous power output of lithium-ion batteries is another benefit. Performance is unaffected while the battery drains. The typical recharge time for lithium-ion batteries is between one and three hours. The added benefit of Li-ion batteries is that they are 30% lighter than alkaline and NiMH batteries, making them ideal for portability.

Although they are not actually a new kind of battery, smart batteries are worth highlighting. The internal circuit boards of smart batteries have chips that enable them to interact with the laptop and track battery life, output voltage, and temperature. Due to their higher efficiency, smart batteries often operate 15% longer and provide a computer with far more precise "fuel gauge" capabilities for calculating how much battery life is available before the next recharge is necessary.

SUMMARY

In this unit, the general idea about the different fossil fuel-based technologies and renewable energy for sustainable power generation is discussed. The data related to these sources is also provided in this unit. The environmental impacts and sustainability in terms of power generation are also described in brief. Further, the role of renewable energy-based technologies in climate change and related issues is also discussed. The possible utilizations of various energy storage systems are also described to bridge the gap when renewable power generation is not available to fulfil the demands.

Short and Long Answer Type Questions

Q1. Describe in brief fossil fuels. What are the advantages and disadvantages of fossil fuels?

Q2. Enlist the various renewable energy sources and give their advantages and disadvantages.

Q3. Describe in detail any four conventional energy sources.

Q4. Discuss in detail the significance of nuclear energy in the present scenario and also give its environmental impact.

Q5. Discuss the functioning of the pumped hydro storage plant.

Q6. What are different renewable energy sources? What is the present status of the development of these resources in India? Present their advantages and disadvantages.

Q7. Why do we consider biomass as the source of energy, especially in developing countries like India?

Q8. Write in brief about biomass resource development.

Q10. What is the potential of biogas development in India?

Q11. Why do we call hydrogen as a secondary source of energy?

Q12. What are the various basic issues in the introduction of hydrogen as an energy source?

Q13. Discuss the potential available of wave energy in India and the world.

Q14. How are the ocean waves produced?

Q15. How the wind energy potential is assessed at a site according to its wind characteristics?

Q16. Write a short note on wind energy development in India and in various leading countries in the world.

Q17. Discuss the past and future trends in energy supply.

Q18. Discuss the role of new energy sources in the context of present-day energy crisis.

Q19. How does wind energy be converted into electrical energy?

Q20. Discuss the future prospects of solar energy use.

Q21.Explain the term co-benefits in terms of energy usage.

Q22. Discuss sustainability and environmental trade-offs of different energy systems.

Q23. Discuss different types of batteries and their relative merits and demerits.

Q24. What is superconducting magnetic energy storage (SMES)? List several advantages offered by SMES when compared to other technologies.

Q25. What are smart batteries? Discuss their advantages.

REFERENCES AND SUGGESTED READINGS

[1] Generation Capacity of India at a Glance, Ministry of Power, Government of India. https://powermin.gov.in/en/content/power-sector-glance-all-india

[2] IEA, "Coal-Fired Power", IEA, Paris, 2021. https://www.iea.org/reports/coal-fired-power

[3] IEA, "Phasing Out Unabated Coal: Current Status and Three Case Studies," IEA, Paris, 2021. https://www.iea.org/reports/phasing-out-unabated-coal-current-status-and-three-case-studies

[4] IEA, "Methane Emissions from Oil and Gas," IEA, Paris, 2021. https://www.iea.org/reports/methane-emissions-from-oil-and-gas

[5] A. K. Jain, and R. Ram, "Shale Gas in India: Challenges and Prospects" NITI Aayog

[6] Ministry of Coal, "National Coal Gasification Mission," Ministry of Coal, Governent of India, 2021. https://coal.gov.in/sites/default/files/ncgm/ncgm21-09-21.pdf

[7] R. Curley, "Energy: Past, Present and Future- Fossil Fuels," Britannica Educational Publishing, 2012.

[8] R. Curley, "Energy: Past, Present and Future- Renewable and Alternative Energy," Britannica Educational Publishing, 2012.

[9] P. Moriarty and D. Honnery, "What is the global potential for renewable energy?," Renewable and Sustainable Energy Reviews, vol. 16, no. 1. 2012. doi: 10.1016/j.rser.2011.07.151.

[10] O. Eseosa and A. E. Ejiroro, "Challenges and Way Forward of Renewables in Developing Energy Economy: Today and Tomorrow," International Journal of Recent Engineering Science, vol. 7, no. 4, 2020, doi: 10.14445/23497157/ijres-v7i4p103.

[11] R. A. Rather et al., "Bioenergy: a foundation to environmental sustainability in a changing global climate scenario," Journal of King Saud University - Science, vol. 34, no. 1, 2022, doi: 10.1016/j.jksus.2021.101734.

[12] P. Duarah, D. Haldar, A. K. Patel, C. di Dong, R. R. Singhania, and M. K. Purkait, "A review on global perspectives of sustainable development in bioenergy generation," Bioresource Technology, vol. 348. 2022. doi: 10.1016/j.biortech.2022.126791.

[13] International Energy Agency, "Renewables 2020 – Analysis - IEA," Iea, 2020.

[14] D. S. Siqueira, J. de Almeida Meystre, M. Q. Hilário, D. H. D. Rocha, G. J. Menon, and R. J. da Silva, "Current perspectives on nuclear energy as a global climate change mitigation option," Mitigation and Adaptation Strategies for Global Change, vol. 24, no. 5, 2019, doi: 10.1007/s11027-018-9829-5.

[15] R. Prăvălie and G. Bandoc, "Nuclear energy: Between global electricity demand, worldwide decarbonisation imperativeness, and planetary environmental implications," Journal of Environmental Management, vol. 209. 2018. doi: 10.1016/j.jenvman.2017.12.043.

[16] A. Kuriqi, A. N. Pinheiro, A. Sordo-Ward, M. D. Bejarano, and L. Garrote, "Ecological impacts of run-of-river hydropower plants—Current status and future prospects on the brink of energy transition," *Renewable and Sustainable Energy Reviews*, vol. 142. 2021. doi: 10.1016/j.rser.2021.110833.

[17] M. Bilgili, H. Bilirgen, A. Ozbek, F. Ekinci, and T. Demirdelen, "The role of hydropower installations for sustainable energy development in Turkey and the world," *Renewable Energy*, vol. 126. 2018. doi: 10.1016/j.renene.2018.03.089.

[18] S. P. Filippov and A. B. Yaroslavtsev, "Hydrogen energy: development prospects and materials," Russian Chemical Reviews, vol. 90, no. 6, 2021, doi: 10.1070/rcr5014.

[19] B. K. Sovacool, "Cobenefits and Trade-Offs of Green and Clean Energy: Evidence from the Academic Literature and Asian Case Studies," ADB Economics Working Paper Series, No. 502, 2016.

****

3 Energy and Environment

UNIT SPECIFICS

In this unit, the following topics have been discussed for basic understanding related to energy and the environment.

- Energy efficiency and conservation
- Clean energy technologies
- Carbon footprint
- Economics of energy
- Influence of the use of energy on the environment, economy, and trade

Here, in this unit the main focus is on energy and its impact on the environment, economy and trade. Growth in the economy resulting in trade expansion and have a direct impact on the environment also. These days the entire focus globally is on the replacement of fossil fuels. Power generation and automobiles have been the major source of carbon emissions. Here in this unit, carbon footprint, clean energy technologies, and linkages between the economy and environment with energy are discussed.

RATIONALE

This unit introduces the concepts of the impacts of the use of energy on the environment, trade, and economy. Energy conservation and efficiency mean a reduction in energy consumption but without making any sacrifice in the quality or quantity of production. Therefore, energy conservation, directly or indirectly, is related to the environment, trade, and economy. Energy efficiency, on the other hand, involves using technology that requires less energy to perform the same function. This unit also discusses how future energy use can be influenced by the economy, environment, trade, and research policy.

PRE-REQUISITES

Basic Knowledge of Environmental Science and Energy

UNIT OUTCOMES

List of outcomes of this unit is as follows

U3-O1: To understand the basic concept of energy efficiency and energy conservation.

U3-O2: Basic idea about the importance of clean energy in sustainable development.

U3-O3: An awareness about carbon footprint and clean energy technologies.

U3-O4: To understand the relationship among energy, environment, trade and economy.

U3-O5: A brief idea about research policy on future energy use.

Unit outcomes	EXPECTED MAPPING WITH COURSE OUTCOMES (1- Weak Correlation; 2- Medium correlation; 3- Strong Correlation)					
	CO-1	**CO-2**	**CO-3**	**CO-4**	**CO-5**	**CO-6**
U3-O1	3	1	3	1	-	3
U3-O2	1	1	3	-	-	3
U3-O3	1	1	3	-	-	2
U3-O4	1	1	3	-	-	2
U3-O5	1	1	3	-	-	2

3.1.1. Environment and its Quality

"*Environmental Quality*" refers to a group of environmental traits that have an impact on people and other living things, whether they are general or specific. It is a gauge of how well an environment meets the needs of one or more species, as well as any human need or goal. Environmental factors, including air, water purity or pollution, noise, and the possible implications these features could have on both mental and physical well-being are considered together with the constructed environment. Whenever we talk about environmental quality, we do not mean the surrounding ecosystems; we also mean the quality of our entire surroundings. The health of the environment (including the plants and animals it sustains) and the impact it has on the physical and emotional well-being of its inhabitants are measured by environmental quality. We can all agree that people's daily environments have an obvious impact on their health, and environmental issues are readily apparent. However, the main challenge is coming up with socially and politically acceptable remedies. We all have a duty to develop solutions to deforestation, pollution, and global warming since environmental issues are caused by social institutions, social conduct, and cultural ideas.

3.1.2. Energy and Environment

The twentieth century saw the emergence of public and scientific knowledge that human activity had negative consequences on the natural environment, including human health and welfare. During this time, industrialization advanced even more quickly than population expansion. These repercussions included increased pollution of air, water, and land by industrial wastes, irreversible loss of native biodiversity due to changes in the estuarine and coastal environment. Infrared-absorbing particles, mostly CO_2 but also nitrous oxide and methane, are more concerning global pollutants that are inexorably building up in the atmosphere and threatening to upset the earth's thermal radiation equilibrium with the sun and space. The majority of scientists now hold the view that this disequilibrium will result in an increase in the average atmospheric surface temperature, which is likely to have unfavorable climatic effects [1]. The continued emission of CO_2 into the atmosphere poses a problem that can only be managed on a global scale because it is formed ineluctably in the combustion of fossil fuels, which generate the majority of the current and anticipated future energy use. Carbon dioxide is also known to accumulate in the atmosphere for centuries.

Governments around the world took steps to reduce the rate of these emissions by requiring technological advancements to pollutant sources because the extent of environmental harm increased in direct proportion to the rate of emission of air and water pollutants, which themselves reflected the rising level of industrial activity. As a result, even while energy and material consumption were rising, ambient air and water pollution levels were progressively declining in the most advanced industrialized countries by the century's conclusion. However, alarming evidence of the cumulative impacts of industrial waste disposal, including the acidification of forest soils, the pollution of marine sediments with urban waste sludge, and the poisoning of aquifers with drainage from hazardous waste dumps, became apparent.

Fossil, nuclear, and hydroelectric fuels are the main energy sources for modern societies. Biomass, wind, geothermal, solar thermal, and photovoltaic energy all make up a minor fraction of the world's current energy output. Fossil fuels, like other mineral reserves, are not evenly distributed around the world; rather, they are concentrated on the borders of continents that historically produced a lot of biomasses. Before they can be used for energy generation, they must be found, removed, and often treated. At present levels of consumption, existing and anticipated deposits would seem to endure for a few decades.

3.1.3. Air Pollution, Stack emissions

Due to the modernization of national businesses, virtually all metropolitan areas in the industrialized world suffered phases of air pollution that were hazardous to human health. Today, this sort of deterioration has expanded to developing nations' metropolitan centers. The combustion of fossil fuels both within and outside of metropolitan areas is the main cause of urban air pollution. The extent of this pollution can reach rural areas some distance from the pollutant sources in large quantities, such that contaminated zones of continental size can even include places where there is no local energy consumption. Although urban smog is severe, it is hypothetically possible to bring it down to safe levels by restricting the emission of the chemical species that damage the atmosphere. The main pollutants make up a relatively tiny percentage of the materials treated, and they may be reduced even more, but at a cost. In developed nations, the expense of reducing urban air pollution represents a very small portion of a country's overall economic output.

The global environment senses an unchecked increase in greenhouse gases, those contaminants that are thought to cause the average surface air temperature to rise and climate change. This is happening while industrialized countries struggle with urban and regional air pollution, with some success, and developing nations lose ground to the intensifying levels of harmful urban air contamination. Contrary to urban pollutants, which are often removed from the environment by precipitation a few days after they are released, greenhouse gases remain in the atmosphere for decades or even centuries [2]. Carbon dioxide, which is emitted when fossil fuels are burnt, is the most prevalent greenhouse gas. It will be extremely challenging to minimize global carbon dioxide emissions while still giving adequate energy to the world's nations for the development of their economies. This is because it is impossible to use all of the energy that fossil fuels have to offer without producing carbon dioxide. Although technology is being developed or is already available that would allow for significant reductions in global carbon dioxide emissions, the cost of implementing such control programs will be far higher than that of reducing urban air pollution.

3.1.4. Cooling tower impacts, Aquatic impacts

Research has shown that the entire energy consumption of process chains is mostly driven by economic activities involving water. In the context of sustainable manufacturing, it is critical to focus on such processes. The so-called water-energy nexus is a term used to explain this reciprocal relationship between water and energy in an industrial setting. Cooling towers (CT) are an integral part of a technical building. These towers are thermodynamically open systems used to moderate the temperature and humidity of the regional environment. These have warm water which is cooled down at ambient air, by transferring heat and evaporated water to the environment. Climate conditions are predicted to have a substantial influence on CT since they are technological equipment that are in close touch with the environment [3]. The essential components of an industrial CT are depicted in the following Fig. 3.1.

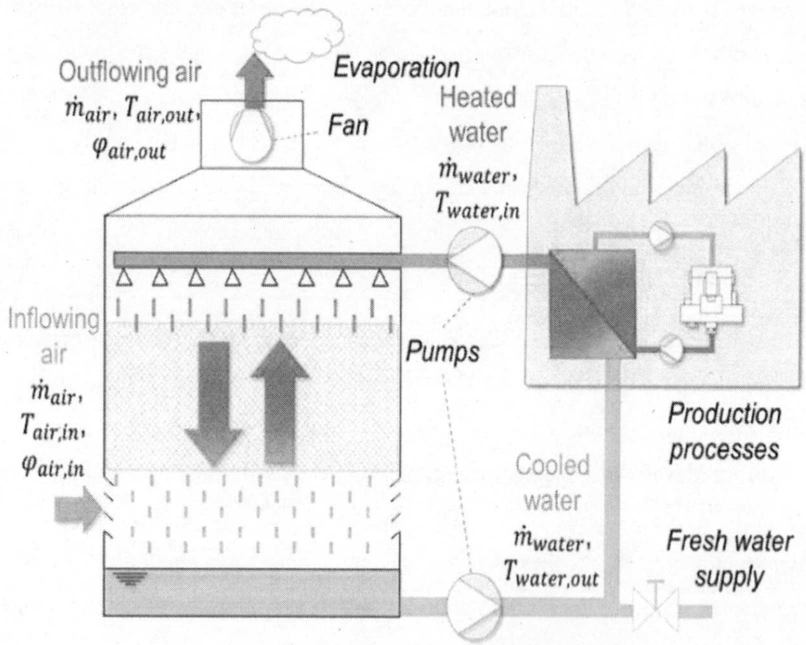

Fig. 3. 1. Elements of industrial cooling towers

A cooling tower is a piece of equipment that uses heat that is extracted from water leaving a condenser and released into the atmosphere to lower the temperature of a water stream. Evaporation is used in cooling towers, where a portion of the water is evaporated into flowing air and then released into the atmosphere. There are different types of cooling systems and towers. They are shown in Fig. 3.2.

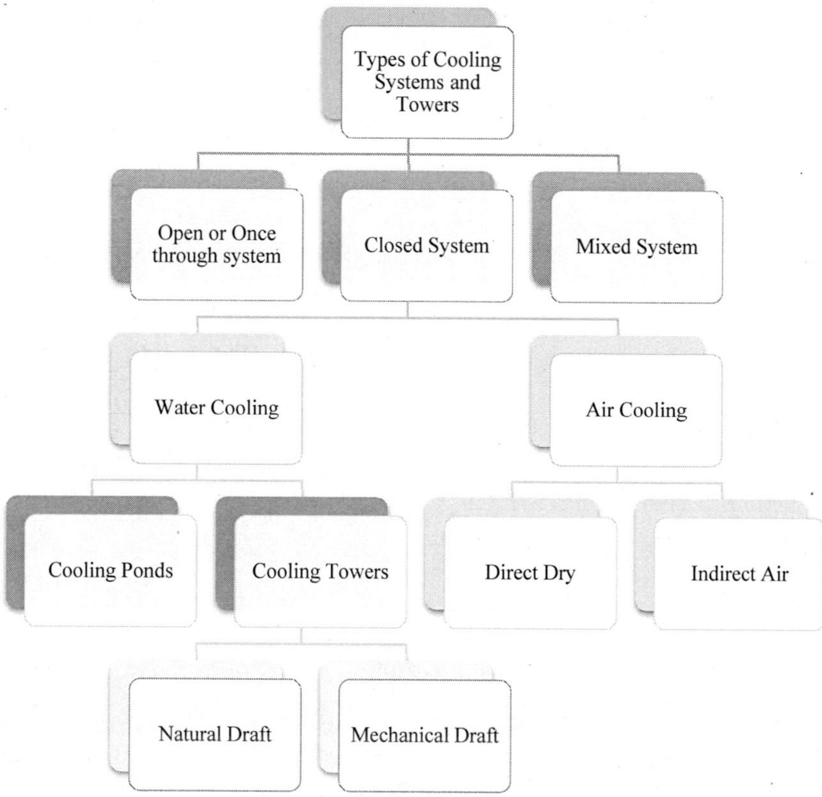

Fig. 3. 2. Different types of cooling systems

3.2. Energy Efficiency and Conservation

Energy conservation (EC) and energy efficiency (EE) are related concepts and frequently overlapping or complementary approaches to minimizing or avoiding energy usage. Energy conversion, consumption, and building material technical performance are all covered under the umbrella term "energy efficiency." Reducing the usage of energy by the consumer is a common component of energy conservation. Installing energy-efficient lighting, for instance, falls under the EE category, but shutting off lights when not in use—either manually or automatically using timers or motion sensor switches—is under the EC category. Table 3.1 presents the difference between energy conservation and efficiency.

Table 3.1. Differences between energy conservation and energy efficiency

Energy Conservation	Energy Efficiency
Energy usage is reduced.	Energy usage is done efficiently and is further minimized if implemented properly.
It brings about a behavioral change to save energy.	This process automates the system and uses less energy to carry out a similar task.
The pre-existing appliances and/ or materials are used in an economical way.	An additional component is required, along with pre-existing or new appliances.
Turning off electrical appliances, using natural light, etc.	Replacing lighting load with LEDs to reduce the consumption of energy.

3.2.1. Motivation for energy conservation

Energy conservation helps in

- Lowering electricity bills
- Improves the rate of return on electrical appliances
- Improves the environment
- Reduces pollution and other greenhouse gases emissions
- Protects the environment
- Promotes alternative low carbon emission sources

3.2.2. Principles of Energy Conservation

The law of conservation of energy states that "***Energy can neither be created nor destroyed but can only be converted from one form to another***". According to the law of conservation of energy, the total energy of an isolated system remains conserved over time.

3.2.3. Energy Conservation in Planning

The energy conservation planning is done in six steps given in Fig. 3.3.

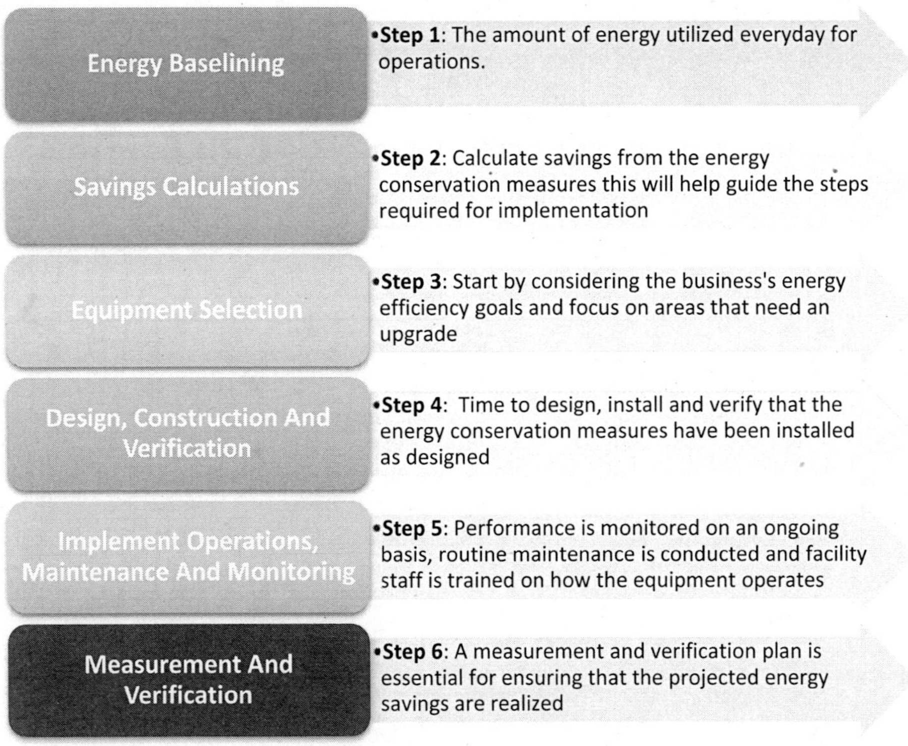

Fig. 3. 3. Energy conservation planning

3.2.4. Energy Conservation in day-to-day life

- Adjust your day-to-day behaviors to turn off devices and appliances when not in use. Purchase devices and appliances which consume less energy.
- Adapt smart power strips: These smart power strips will help to cut down on phantom-load costs and save energy.
- Designing a building to use maximum daylight.
- Perform regular home energy audits.
- Refrigerators are one of the main appliances that consume power. Keep the setting of the refrigerator low to save energy.
- Install LED bulbs to save energy. Regular incandescent and CFL bulbs consume more energy than LEDs.
- Proper insulation of the rooms or premises to enhance cooling and fixing air leaks.
- Clean or replace air filters as recommended. Cleaning or replacing air filters in air conditioners improves efficiency and consumes less energy.

- Operate washing machines in a full load to get the most energy-saving use from each run cycle.
- Using a laptop instead of a desktop computer can save considerable energy.
- Install water-saver showerheads to help with conserving hot water and save power.
- Use a slow cooker, toaster oven, or microwave oven over a conventional oven. Also, use utensils made of ceramic and glass.
- Cycling is the best way to save fuel.
- Walking instead of driving also saves energy.
- Skip the dryer on breezy day and dry clothes on the clothesline.

3.3. Introduction to clean energy technologies and their importance in sustainable development

Clean energy is defined as energy generated using renewable, non-polluting, zero-emission resources and includes energy which is conserved through energy-saving practices. Clean energy and green or renewable energy sources have some overlap, but they are not the same thing entirely. It is important to grasp what it really implies in order to recognize the difference.

The power produced from resources that are continuously renewed is known as renewable energy. Unlike fossil energy and gas, these renewable energy sources, which include wind and solar energy, won't run out. Despite the fact that most green forms of energy are renewable, not all green energy comes from renewable sources. For instance, hydropower is a renewable resource, yet some might suggest that it's not environmentally friendly due to the deforestation and industrialization associated with the construction of hydro dams. Green energy and renewable energy, including solar and wind energy, combine to provide the ideal clean energy combination.

The different types of energy and their associativity with different sources can be seen as follows:

- ***Clean energy*** is equivalent to ***clean air***
- ***Green energy*** is equivalent to ***natural sources***
- ***Renewable energy*** is equivalent to ***recyclable sources***

The advantages of clean energy as a component of the future world's energy are its most significant feature. Clean, renewable resources not only protect the planet's mineral resources but also lessen the likelihood of ecological catastrophes like fuel spills or the issues brought on by natural gas leaks. It is feasible to generate dependable power supplies to improve energy security, ensuring there is enough to fulfill our demands, through diverse power plants employing different energy sources and fuel diversity.

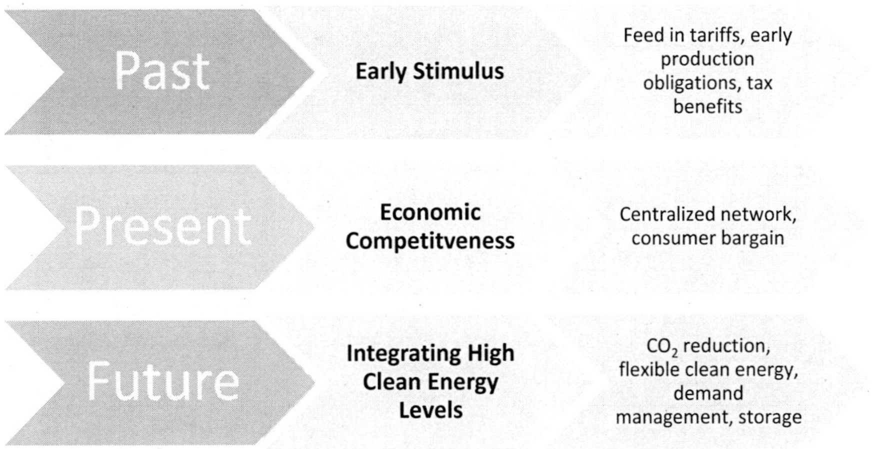

Fig. 3. 4. Evolution of clean energy policy and markets

In recent decades, clean energy technology has grown very quickly. Although there has been substantial progress, many want to see it accelerate in order to meet the many issues facing the world today, such as national defense, alleviating poverty, environmental destruction, dealing with climate change, and wealth creation. The adoption of wind and solar technologies was ramped up by early clean energy regulations, including feed-in tariffs, rebate/incentive programs, and renewable energy mandates [4]. Early deployment that was successfully allowed for cost savings and more technological advancement. Policymakers were encouraged to raise clean energy goals as a result, and they are currently developing market reforms to include significant amounts of wind and solar energy into the power grid. The transportation or heating and cooling sectors, however, have not benefited as much from early stimulus initiatives as they could have, and more aggressive follow-up policies and market planning have not yet followed. A lot of power plants are being forced to retire early as a result of competition from cheap sustainable energy. In order to balance renewable energy supplies that are getting farther away from energy consumption, markets run by centralized network operators are expanding as a result of the more decentralized structure of the energy supply. Conventional capital-intensive energy balancing areas are being replaced by this. Low-cost renewable energy technologies are also boosting energy customers' bargaining power as they organize and encourage fresh investment through direct contracts with clean energy suppliers.

Target 7.1

By 2030, ensure universal access to affordable, reliable and modern energy services

Target 7.2

By 2030, increase substantially the share of renewable energy in the global energy mix

Target 7.3

By 2030, double the global rate of improvement in energy efficiency

Target 7.a

By 2030, enhance international cooperation to facilitate access to clean energy research and technology, including renewable energy, energy efficiency and advanced and cleaner fossil-fuel technology, and promote investment in energy infrastructure and clean energy technology

Target 7.b

By 2030, expand infrastructure and upgrade technology for supplying modern and sustainable energy services for all in developing countries, in particular least developed countries, small island developing States, and land-locked developing countries, in accordance with their respective programs of support

Fig. 3. 5. Sustainable development goals linked with clean energy technologies

(For more details, please scan)

In order to handle the dynamic nature of a high-clean energy system, clean energy enabling methods and resources will be crucial in the future. In order to handle the unpredictable supply caused by high wind and solar penetration, this includes obtaining more flexibility for the energy system through generation resources, demand, and storage. By consolidating and enlarging the boundaries of the power market, early levels of wind and solar fluctuation are being controlled. Another very inexpensive source of adaptability for the penetration of renewable energy is to rely on the scalability of current conventional and clean energy facilities. Storage solutions and better demand-side technology will be required as renewable energy usage increases. For technology researchers, these domains provide intriguing opportunities to facilitate and quicken the move toward the future. The sustainable development goals associated with clean energy and the environment are given in Fig. 3.5.

3.4. Carbon Footprint

It is the sum of greenhouse gas emissions and removals in a product system, expressed as CO_2 equivalents and based on a life cycle assessment using the single impact category of climate change. The concept of carbon footprint can be traced back to 1996 when it was under the subset of "ecological footprint" [5]. The usage of carbon footprint eventually spread organically, albeit in a slightly different form, as the subject of global warming gained importance on the international environmental agenda. The idea of a carbon footprint has been around for a while, but it is often referred to as the possibility for life cycle impact categories to indicate global warming. As a result, the current version of the carbon footprint may be seen as a hybrid, taking its name from the phrase "ecological footprint" and theoretically serving as a possible indication of global warming. Despite the current nexus around it, there aren't many studies that measure carbon footprint in terms of world hectares.

The choice between direct and embedded emissions is not consistently applied. Direct emissions are those produced immediately as a process moves forward. An industrial boiler that burns gasoline produces direct emissions, such as the CO_2 that is produced during combustion. On the other hand, no direct emissions will be seen in a boiler that is heated by electricity. The quantity of CO_2 emitted during the production and transmission of the units of energy used in the boiler, however, is known as the embodied or secondary emission if it came from a thermal power station where the boiler's electricity was produced. Since it becomes difficult to account for all potential emissions, most studies simply provide direct or first order indirect emissions.

3.4.1. Calculation of Carbon footprint

Some countries have created their own carbon footprints accounting guidelines, such as the Department of food and rural affair (DEFRA) and Carbon trust in the UK, Environment Protection Agency (EPA) in the US. The USA also has California Climate Registry and World Wildlife Fund Climate Servers. These bodies themselves have formulated their own methodologies based on the guidelines provided by World Resource Institute (WRI) and Intergovernmental Panel on Climate Change (IPCC).

Table 3.2. Global warming potential of some GHGs (please scan for more details)

GHG	Global warming potential
CO_2	1
CH_4	25
N_2O	298

The process for calculating carbon footprints is based on life cycle assessment (LCA) methodology and life cycle thinking. The amount of greenhouse gases (GHGs) created throughout the course of a product's life cycle is referred to as its "carbon footprint." Using 100-year global warming potentials, greenhouse gas emissions are translated into carbon dioxide equivalents. All of the greenhouse gas emissions stated by the IPCC are accounted for in the carbon footprint scenario estimates in this document.

The quantity of GHGs released, withdrawn, or embodied over the life cycle of the product must be calculated and summed in order to calculate carbon footprint. The term "life cycle" refers to the entire process of creating a product, from the procurement of raw materials through final packaging, distribution, consumption, and usage to the last steps of disposal. Cradle-to-grave analyses are another name for life cycle analysis. With regard to the production of air pollutants, water use and wastewater generation, energy consumption, GHG emissions, or any other comparable parameter of interest and cost-benefit measures, life cycle assessment (LCA) generates a full picture of inputs and outputs. This analysis is frequently referred to as an environmental LCA. Key steps for estimating carbon footprint are presented in Fig. 3.6.

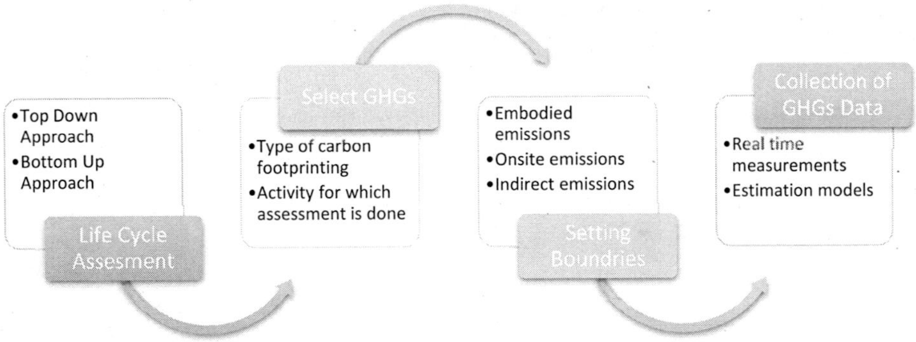

Fig. 3. 6. Stages of performing carbon foot printing

3.4.2. Importance of Carbon foot printing

A company's energy and raw material consumption inefficiency or inadequacies can be found via a carbon footprint analysis. This information is crucial for a company to guarantee that knowledgeable decisions can be taken to overcome such adversities in light of rising electricity prices and expected shortages of water and other natural resources. Carbon foot printing does not need to be resource-expensive since it leverages information that has already been gathered by a company, such as energy bills and trip expenditure claims. However, improving a business's access to this crucial demographic is more than just a checkbox exercise. As was already said, it aids a company in finding methods to save costs and risks.

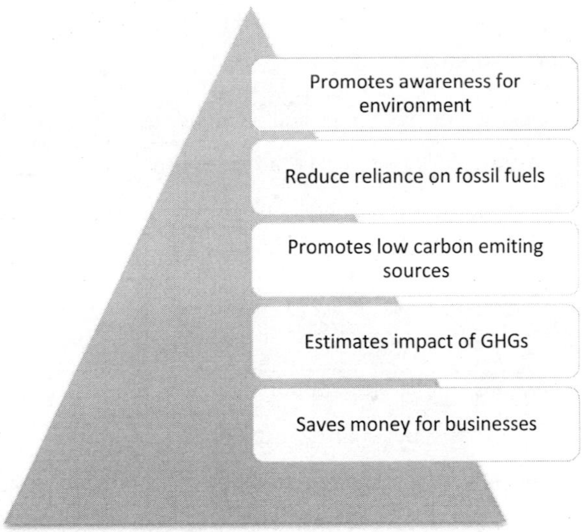

Fig. 3. 7. Benefits from carbon foot printing

Some examples of reducing carbon footprint are listed below:

- In general, single-family homes utilize more electricity per inhabitant than multifamily housing, including apartment complexes. However, there are measures you can undertake to lower the quantity of energy your residence uses and so lessen your carbon impact, regardless of where you reside.
- A smart thermostat dynamically adapts the temperature in your house according to the time of day. Your carbon footprint decreases when your apartment's energy system is operating well because it wastes less energy.
- Appliances with the Energy Star label are more energy efficient. To save energy, money, and the environment when buying new appliances like a fridge, gas cooker, toaster, or air conditioning system, seek for the Energy Star label.
- Composting benefits the environment by preventing food and yard waste from entering landfills and by promoting organic gardening techniques. Your carbon impact is lower if you compost at home.
- If you often consume organic food, you have a reduced carbon footprint because raising natural foods is less stressful on the planet.
- Cars that receive regular maintenance use less energy. Make sure the air filter is changed frequently and that the tires are constantly filled appropriately.
- Recycling is good for the environment because it keeps excess waste out of landfills and trash incinerators. If you recycle on a regular basis, your carbon footprint is smaller.

3.5. Introduction to the economics of energy

The study of how people use energy resources and commodities, as well as the effects of such use, is known as energy economics. Energy is the ability to perform labor, such as lifting, accelerating, or heating material, in the language of physical science. According to economic language, "energy" refers to any goods and resources that include a considerable quantity of physical energy and may be used to accomplish work. Energy services for human activities like lighting, space heating, water heating, cooking, motive power, and electronic activity can be provided by energy commodities like gasoline, diesel fuel, natural gas, propane, coal, or electricity. Energy commodities can be produced by harvesting energy resources such as crude oil, natural gas, coal, biomass, hydro, uranium, wind, sunshine, or geothermal deposits.

Energy economics examines the factors that motivate economic actors—firms, people, and government agencies, providing energy resources, transform those resources into various forms of usable energy, transfer those resources to consumers, use those energy sources, and dispose of the waste products. It examines how these activities are impacted by various market and regulatory arrangements, as well as their effects on economic inequality and the environment. It examines issues that lead away from economic efficiency in the provision and use of energy commodities and resources.

3.5.1. Cost analysis of production and consumption

The overall expense paid by a company to create a particular amount of a good or provide a service is referred to as the cost of production. Costs associated with production may include labor, raw materials, or consumable supplies. The costs paid to acquire the components of production—such as labor, land, and capital—necessary for a product's manufacture are referred to as the cost of production in economics. The costs linked to various electricity production strategies can be broken down into three broad categories: 1) wholesale expenses, or all expenses incurred by utilities in the purchase and distribution of electricity to customer; 2) retail costs incurred by purchasers; and 3) external costs, or externalities imposed on society.

Table 3.3 define some of the cost metrics that are readily used.

Table 3.3. Cost metrics associated with electricity production and consumption

Name	Description
Levelized Cost of Electricity (LCOE)	Minimum constant price at which electricity must be sold in order to break even over the lifetime of the project
Levelized Cost of Storage (LCOS)	Analogous to LCOE, but applied to energy storage technologies such as batteries
Levelized Avoided Cost of Electricity (LACE)	This metric considers some of the shortcoming of LCOE as it takes into account the dispatchability of a resource as well as the existing energy mix in a region
Value Adjusted Levelized cost of Electricity (VALCOE)	This metric is introduced by International Energy Agency, and includes both the cost of the electricity and the value to the electricity system
Cost Factors	
Capital Cost	Minimum for gas and oil power stations, moderate for PV and wind turbines and high for coal based thermal plants

Fuel Costs	High for fossil fuels, low for nuclear power and negligible for renewable sources
Operations and Maintenance Cost	O&M costs include marginal costs of fuel, maintenance, operation, waste storage, and decommissioning for an electricity generation facility.

The major finding of the research continues to be the growing competitiveness of low-carbon technologies for energy generation. This is true for flexible low-carbon producers like hydropower and nuclear energy in addition to intermittent renewable sources like wind and solar PV. Coal that has not been treated has lost its competitiveness, even at a low carbon cost of USD 30 per tonne of CO_2 [6]. Due to the extremely low gas costs, gas-fired power generation is still competitive in several regions, particularly OECD North America. To be competitive, CCUS would need carbon prices that are far higher than those seen in the majority of markets now.

Some of the key terms associated with the economics of energy are explained in Fig. 3.8.

Ramp Rate

- This variable influences how quickly the plant can increase or decrease power output, in [MW/h] or in [% of capacity per unit time]

Ramp time

- The amount of time it takes from the moment a generator is turned on to the moment it can start providing energy to the grid at its lower operating limit (see below), in [h]

Capacity

- The maximum output of a plant, in [MW]

Lower Operating Limit (LOL)

- The minimum amount of power a plant can generate once it is turned on, in [MW]

Minimum Run Time

- The shortest amount of time a plant can operate once it is turned on, in [h].

No-Load Cost

- The cost of turning the plant on, but keeping it "spinning," ready to increase power output, in [$/MWh]

Start-up and Shut-down Costs

- These are the costs involved in turning the plant on and off, in [$/MWh].

Fig. 3. 8. Key term associated with energy economics

3.5.2. Linkages Between Economic and Environmental Outcomes

The distribution and use of environmental resources are taken into account in the context of their financial cost and benefit. Economics is used to examine supply and demand, gains and losses, and the equilibrium of environmental resources. Numerous environmental ideas have emerged as a result of the incorporation of economic theories. Input-output models, environmental cost analysis, environmental policy, environmental pricing, environmental budgeting, environmental fiscal analysis, and environmental resource planning are a few of these. Natural resource scarcity is a major issue for emerging nations. In the setting of finite natural resources, economists can provide environmental analysis guidance to satisfy desires to the greatest extent possible. Environmentalists can use economics to help them determine how to maximize benefits or minimize losses. Pollution issues can be described using economic language.

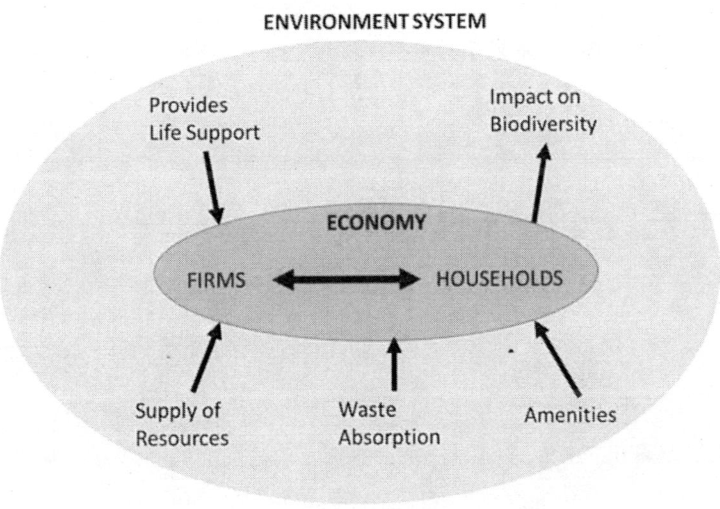

Fig. 3. 9. Economy Environment Linkages

Economic issues have environmental causes, while environmental concerns have economic causes. There are environmental solutions for economic issues as well as economical solutions for environmental issues. The same holds true for economic theories, which are necessary for both environment theories and economic theories. This association between economy and environment is given in Fig. 3.9. Water pollution and air pollution are mostly caused by home and industrial waste. Land contamination is caused when contaminated water is absorbed by the land. This can be resolved through economics. According to economics, public goods, including air, land, water, rivers, and oceans, transmit pollution. Controlling these contaminated public commodities is necessary. Economics demonstrates a number of ways and means to address environmental issues through input-output analysis, cost-benefit analysis, pollution taxes, and environmental subsidies. The majority of environmental issues are caused by people, and economics offers solutions. The overuse of natural resources should be avoided. We should establish some fundamental guidelines for the usage of natural resources.

The Empirical Kuznet Curve (EKC) illustrates the empirical trend that, at relatively low average income per capita values, pollution level (and intensity) rises initially with growing income, reaches a maximum, and then declines. It is shown in Fig. 3.10. As a result, the EKC demonstrates that there is an inverse U-shaped link between pollution and economic development.

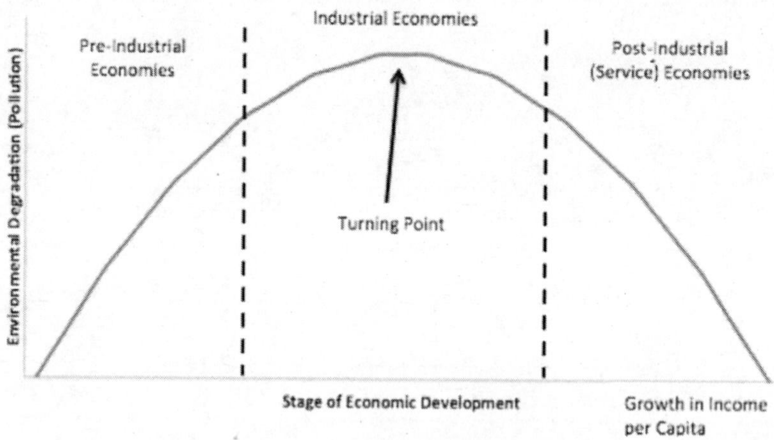

Fig. 3. 10. The Empirical Kuznet Curve [7]

Particularly in emerging nations with a rapidly expanding population and widespread poverty, the clash between economic progress and the environment is more acute. The emerging nations are attempting to strike a compromise between their desire for rapid economic expansion and environmental concerns related to maintaining their natural resources. Environmental degradation is a result of the adoption of a development plan focused on industrialization, energy-intensive technologies, and biochemically based agriculture technology. The idea of "sustainable development" has been proposed in light of the connections between the natural environment and development. Today's sustainable development aspires to enhance human well-being on a broad scale, encompassing economic prosperity, social fairness, and environmental conservation. Additionally, it mentions cultural, social, political, and economic factors.

3.5.3. Impacts on Energy Use by Economic, Trade, and Research Policy

Energy consumption and economic trade are related, although the intensity of the relationship varies among locations and economic development phases. The relationship between economic trade and energy consumption is significantly influenced by the level of economic development and the standard of living of the population in a particular area. Advanced economies with high living standards require a lot of energy per person, yet they also frequently have constant or gradual changes in per capita energy usage. Modern appliances and motorized personal transportation devices are widely used in industrialized nations. When money is spent on energy-intensive products, it frequently entails replacing the outdated capital stock with new equipment. Due to the fluctuation in global GDP growth, some nations consistently lead their respective regions in that regard. The patterns and impact of these nations, especially China, India, Brazil, and South Africa, will be crucial in determining not only the electricity, financial, and environmental issues mitigation trends of the developing world, but also those of the remainder of the world. These nations' massive energy infrastructure investment requirements over the coming decades will open up a rare window of opportunity for the development of green, low-carbon energy. They will also still have to contend with the difficulties of maintaining economic development and eradicating poverty.

In addition to making a significant number of locations available for frequently contentious renewable power facilities and yearly training a large number of energy staff at all levels, the route to environmental stewardship will require nearly US$2 trillion in investment, primarily from the private sector. Generally speaking, innovation-supporting government institutions are unprepared to handle this task. Large, risk-averse firms with vested interests in maintaining their cozy status quo and history of underinvesting in innovation are likewise dominant in the energy industry. Current studies to support climate-friendly development have their attention focused on the growing significance of investment flows to encourage technology. The private sector, whether through company R&D, venture capital or asset finance agreements, or money obtained in public markets, accounts for the bulk of global investment and technological dissemination. Accordingly, studies evaluating technology transfer and dissemination in developing nations have found that trade openness is an essential condition for effective transfer. Given the significance of the R&D effort for creating new low-emission technologies and the observed decline in activity, it is likely that the government will continue to play a role in this field. Governments have a three-fold role that goes beyond energy research and development. First, they support knowledge creation in general (through education, international science and technology cooperation, and information exchange); second, they support basic and applied energy technology R&D through direct public R&D expenditures; third, they create and maintain the right incentives for private sector R&D. Finally, they support market deployment incentives and compensation.

Summary:

In this unit, a relationship between energy, environment, trade, and economy and the impacts of the use of energy on these three important components, environment, trade, and economy, are discussed. The importance of energy efficiency and energy conservation are also explained. Carbon emissions are the major source of pollution in the environment. Carbon foot printing and clean energy technologies are also discussed in length. The significance of the R&D effort for creating new low-emission technologies and the role of the government to support R&D, venture capital or asset finance agreements, or money obtained in public markets, which accounts for the bulk of global investment and technological dissemination are also discussed in this unit.

Short and Long Answer Type Questions:

1. What is the necessity to limit carbon emissions to environment?
2. Name the various stack emissions.
3. Explain Empirical Kuznet Curve (EKC).
4. What are particulates? What are their harmful effects?
5. Name the various cooling tower impacts. What are the harmful effects of cooling tower emissions?
6. Name some energy conservation measures in industries.
7. How can energy conservation be done in agriculture?
8. Discuss some energy conservation measures for household and commercial sectors.
9. How do CO_2 emissions affect environment?
10. How does energy supply affect international relations?
11. What is the greenhouse effect?
12. What are the various environmental problems associated with energy supply and use?
13. What are the various efforts which countries must undertake for sustainable energy development?
14. Write a note on linkages between economic and environmental Outcomes
15. What is economics of energy? Explain cost analysis of production and consumption

References and suggested readings:

1. A. R. Hoffman, “Water, Energy and Environment – A Primer,” IWA Publishing, London, 2019.
2. J. A. Fay, D. S. Golomb, “Energy and The Environment,” Oxford University Press, New York, 2002.
3. C. Schulz, B. Raabe, C. Herrmann, S. Thiede, “Environmental Impacts of Cooling Tower Operations-The Influence of Regional Conditions on Energy and Water emands,” Life Cycle Engineering Conference, vol. 69, pp. 277-282, 2018.
4. S. Dahlke, J. Sterling, C. Meehan, “Policy and market drivers for advancing clean energy,” Advances in Clean Energy Technologies, Academic Press, Elsevier, pp. 451-485, 2021.
5. M. Wackernagel, W.E. Rees, “Our ecological footprint: Reducing human impact on the earth, Gabriola Island,” New Society Publishers, 1996.
6. IEA, “Projected Costs of Generating Electricity,” 2020, IEA, Paris https://www.iea.org/reports/projected-costs-of-generating-electricity-2020
7. G.M. Grossman, A.B. Krueger, “Economic Growth and the Environment”, Quarterly Journal of Economics, vol. 110, pp. 353-377, 1996.

****

4 Civil Engineering Projects Connected with Energy Sources

UNIT SPECIFICS

In this unit, the following topics have been discussed related to civil engineering projects connected with energy sources:

- Technologies used to develop infrastructure to build power plants
- Coal Mining Technologies
- Oil Exploration Offshore Platforms
- Underground and under-sea oil pipelines
- Solar Chimney Project
- Coastal Installations for tidal power plants
- Wind towers
- Civil infrastructure for building hydropower plants

RATIONALE

This unit introduces the infrastructure and related civil work required to build different types of power plants. The use of technologies has made it easy, efficient, and convenient for workers to develop energy infrastructure underground, under the sea, and over the sea. Different technologies used in coal mining, oil exploration, offshore platforms, under–ground and undersea oil pipe lines have been discussed and presented here.

PRE-REQUISITES

Basic knowledge of energy related projects & infrastructure used

UNIT OUTCOMES

List of outcomes of this unit is as follows

U4-O1: To know about the infrastructure & technologies used in the supply of different energy sources.

U4-O2: To know about the coal mining technologies and construction of a solar chimney.

U4-O3: To be familiar with oil exploration technologies.

U4-O4: To know about windmill towers and related infrastructure.

U4-O5: To understand the civil work required in building a hydro power plant.

Unit outcomes	EXPECTED MAPPING WITH COURSE OUTCOMES (1- Weak Correlation; 2- Medium correlation; 3- Strong Correlation)					
	CO-1	CO-2	CO-3	CO-4	CO-5	CO-6
U4-O1	1	1	2	3	-	-
U4-O2	1	1	1	3	-	-
U4-O3	-	1	1	3	-	-
U4-O4	-	1	1	3	-	-
U4-O5	-	1	2	3	-	-

4.0 Coal Mining Technologies

Introduction

Coal is the most accessible fuel in India. Coal-fired power stations fulfill 60% of India's electricity needs. After China, India is the world's second–largest producer of coal. Coal is derived from organic compounds such as wood. When vast parts of the forest are hidden in deep sedimentary rocks, the wood burns and decomposes as a result of hot air from the ground and pressure from the outside environment. The process produces coal but takes generations to accomplish. Based on its carbon content, coal is categorized as lignite or brown coal, bituminous coal or black coal, anthracite, and peat. Each category of coal has a unique collection of physical properties that are primarily influenced by humidity, high volatility content, and carbon content. Anthracite is the finest quality coal, with the maximum calorie content and a carbon content ranging from 80 to 95%. It has a slow blue flame and is found in relatively small amounts in the regions of Jammu and Kashmir. Bituminous coal has a low water content, 60 to 80% carbon content, and an increased caloric content. Bituminous reserves can be found in Jharkhand, West Bengal, Odisha, Chhattisgarh, and Madhya Pradesh. Lignite contains 40 to 55% carbon and is usually brown in color with a high-water content, producing fumes when combusted. Lignite deposits can be found in Rajasthan, Assam, and Tamil Nadu. Peat is the initial stage of the wood-to-coal conversion process, with low-calorie content and less than 40% carbon content.

4.1 Coal Mining Technologies

Mining activities and related activities provide a significant amount of energy. As earlier as the 12th century AD, humans extracted coal by choosing and scraping and utilized it for warming, preparing food, rituals, and ceremonies. Coal was first employed commercially in traditional crafts in the 14th century. Surface reserves have been mined on a small level for a long time. In the early 1800s, coal was mined in America, and commercial mining began around 1730 in Midlothian, Virginia. Coal mining in India began in 1774 in coal mines in Bengal. Jharia, Raniganj, Singareni, Talcher, Neyveli, Singrauli, Nagpur, and Chandrapur are the major coal domains in India. Jharia coal field (JCF) is India's foremost storage facility of prime coking coal, feeding a large portion of our industrial production. JCF is primarily a producer of bituminous coal. The JCF currently has about 35 large underground and opencast mines.

In the 1880s, coal-cutting machines were created. Prior to this discovery, coal was extracted from the ground using a pick and shovel. Surface mining was done with steam shovels intended for coal mining by 1912.

Coal mining has evolved significantly over the years, from men digging a tunnel, and individually retrieving coal on carts to large open–cut and long–wall mines.

Surface mining and underground mining are the two main types of coal seam mining. The classification of coal mining methods is displayed in Fig. 4.1.

4.1.1 Classification of coal mining methods:

Fig. 4.1 Categorization of coal mining methods

Surface Mining Method:

Surface mining is responsible for 40 percent of the world's coal supply. When the coal seam is close to the surface and the depth is less than 50 meters, this method is used. As a result of the fact that all coal seams are mined. This method can recover a greater percentage of the coal deposit than underground mining. In fact, this method can recover at least 90 percent of the coal.

Explosives are first used to break up the overburden of soil and rock, and then either draglines or shovels and trucks are used to remove it from the site. After the coal seam has been uncovered, it is drilled, fractured, and then mined in strips using a systematic approach. Large trucks or conveyors are utilized to load the coal for transfer to either the coal preparation facility or directly to the location where it will be used.

Huge opencast mines may cover an area that is several square kilometers and employ extremely giant pieces of equipment, such as draglines, power shovels, large trucks, bucket–wheel excavators, and conveyor systems. The various steps involved in the coal mining process are shown in Fig. 4.2.

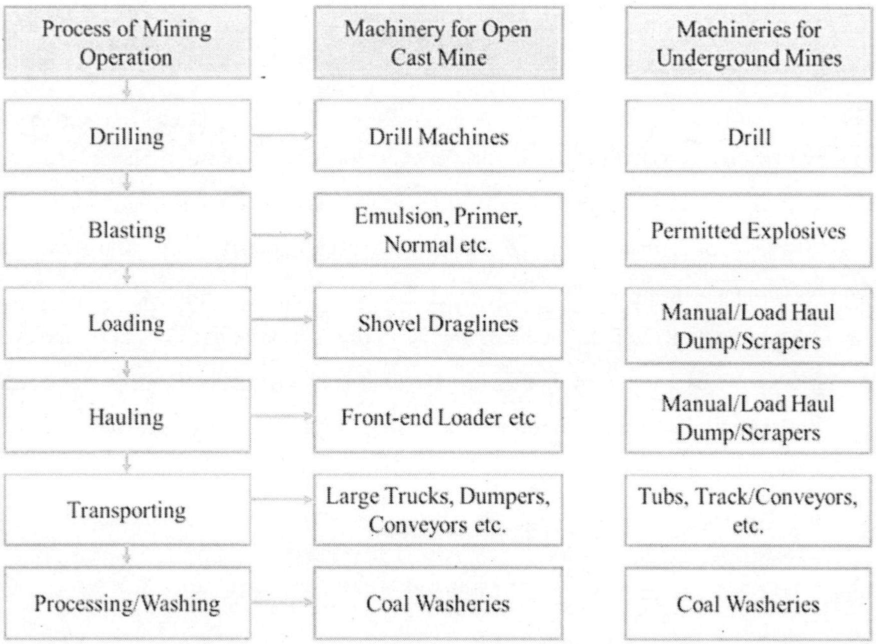

Fig. 4.2. Layout for coal mining processes

(a) Strip Mining:

Open–cut mining is another name for strip mining. It is often used in coal and mineral mines that are close to the surface and not very deep. It uses very big machines like draglines, shovels, and dump trucks to move soil, dig up coal or other minerals, and move them to a place where they can be processed. This type of large-scale mining works best with thick horizontal coal seams that are not too far below the surface. In strip mining, the resource or ore is taken out by digging, drilling, and/or blowing it up. Often, there is an "overburden" or waste rock that must be removed before the real resource can be mined. The ratio of overburden to coal can be as low as 1:12 up to a depth of 50 feet or as high as 1:15 when the overburden is about 90 feet deep. If the overburden is only a few feet thick and the conditions are right, bulldozers can be used to strip the land. Open-cut coal mines are cleaned up in stages over the course of their life. Rehabilitation is a requirement of both the mining lease and the approval for development. It is often done in the USA.

(b) Slice Mining: Slicing is commonly applied to that method of mining in which the ore is extracted by a series of timbered slices, beginning at the capping; the slice is caved, bringing down the capping to the floor of the slice; the succeeding lower slices are mined up to the overlying mat or gob, consisting of an accumulation of timber from the upper slices and broken capping, which follows the mining progressively downward. The timbers are blasted out, bringing the capping down to the floor, succeeding slices being driven alongside those already caved.

(c) Horizon Mining: It is a way to mine coal that has more than one level. Level roads are driven into the rock to open the coal seams. This method can be useful in places where there are a lot of rocks and a lot of coal seams are known to be. This method is used in places where there has been a lot of damage and the coal seams have been folded and faulted. It is often done in Germany, France, and Belgium.

Deep Underground Mining: Underground mining is used to get 60% of all the coal in the world. Most mines use the Bord and Pillar method, the long wall method, or the short wall method. In the Bord and Pillar method of coal seam mining, a series of narrow headings are driven into the seams in a line. These headings are linked by cross-headings to make pillars that can be taken out in whole or in part later. The pillars should be square, but sometimes they are rectangular or rhombus-shaped, and the galleries around the pillars are always square. The method works best on flat coal seams that are only 1.8 to 3 m thick and are close to the surface.

4.1.2 New Technologies used for Coal Mining

With so many new advancements in technology in mining companies, it is no surprising fact that this sector is expected to keep expanding and enhancing its mechanisms. Many current mining technologies are anticipated to have an ongoing impact on the sector in the coming years, including:

- Automation
- Artificial intelligence
- 3D modeling
- GIS.
- Drones.
- Virtual reality.

In relation to all these active mining alternatives, there are numerous developments that will frame the mining industry's future.

Digitalization and Big Data

To improve its efficiency, the mining sector will continue to use data. Industries will be able to analyze data using automated techniques to make the most educated decisions about the process of production, schedule activities, mine situations, and inventory management, to name a few. Greater insight leads to better analytics, safety, and operational efficiency.

Sustainability and Waste Prevention

Sustainable development is one tendency that is intended to have a significant impact on mining methods in the future. Waste recycling and fuel efficiency are crucial aspects of the sustainability movement.

Future mining machines and techniques that are used optimally can significantly reduce the amount of water and emission levels. Furthermore, mining companies will work to minimize their environmental footprint in sensitive or environmentally challenging areas.

Internet of Things

The Internet of Things (IoT) is a collection of physical items such as hardware, sensors, and data sources. All networking devices that are integrated can communicate with one another, exchange data, and act in concert. They facilitate low-cost network establishment in the mining industry while collecting actual information from mining machines and equipment. Industries can use this information to aid judgment, perform integrated control and planning, and consider different operational units. The application of IoT technology in the coal mining process is illustrated in Fig.4.3.

4.2 Oil exploration offshore platforms

An oil platform, also called an offshore platform, is a huge building with equipment to obtain oil and natural gas from the foundation of rocks under the sea floor and process them. The personnel has their rooms on the premises of separate platforms. The platform is attached to the main platform by means of a bridge. Many oil platforms operate on the inland shelf, but they can be operated in ponds, close to shore waters, and inland oceans. The platform may be set on the ocean floor in some circumstances. Also, there are places for storing refined oil in many cases. Flow lines and umbilical connections can also be used to link subsea wells that

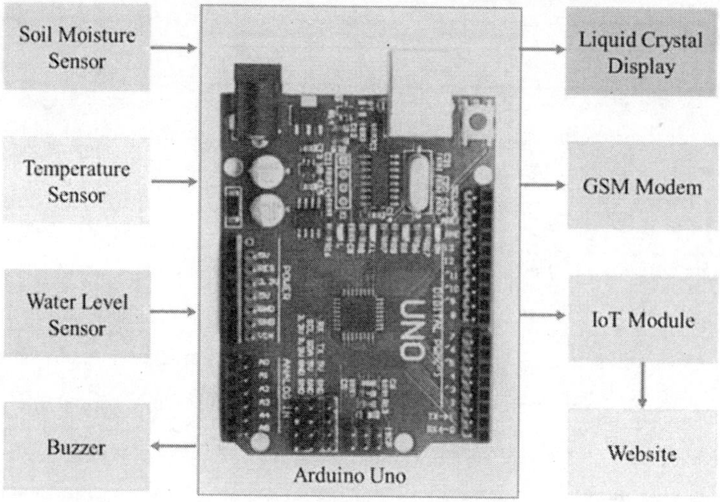

Fig. 4.3. IoT application in coal mining

away from a platform. These subsea solutions can include one or more subsea wells or one or more manifold centers for multiple wells.

Both the hydrocarbons that are made and the materials that are used during offshore drilling are bad for the environment. The ongoing US offshore drilling debate is one example of a conflict.

Types of offshore platforms

1) Fixed platforms

The legs of these platforms are made of concrete, steel, or both. They are anchored directly to the seabed and hold up the deck, which has space for drilling rigs, production facilities, and living quarters for the crew. Because these platforms cannot move, they are made for very long-term use (for instance the Hibernia platform). Structures like steel jackets, concrete caissons, floating steel, and even floating concrete are used. Steel jackets are made of tubular steel members and are usually piled into the seabed.

2) Compliant towers

These platforms are made up of thin, flexible towers and a pile foundation that holds up a standard deck for drilling and production. These are made to withstand large sideways movements and pressure exerted on them. They are generally employed in water depths between 370 and 910 meters.

3) Semi-submersible platform

Semi-submersible platforms have casings that are light enough to float but heavy enough to keep the structure standing straight. Semi-submersible platforms can be moved from one place to another, and their weight can be changed by changing how much water is in their buoyancy tanks. During drilling and/or production, they are usually kept in place with a combination of chain, wire rope, polyester rope, or both. They can also be kept in place with dynamic positioning. Semi-submersibles can go into the water as deep as 60 to 6,000 meters (200 to 20,000 ft).

4) Jack-up drilling rigs

Jack-up Mobile Drilling Units, also called "jack-ups," are rigs that can be lifted above the water using legs that can be lowered like jacks. Most of the time, these Mobile Offshore Drilling Units are employed in water up to the depth of 120 m, but few models can dive to depths of 170 meters. They must move from one position to another before anchoring themselves by placing their legs on the ocean floor and using a rack and pinion gear system on each leg.

5) Drillships

A drillship is a ship that is linked with equipment to be drilled. It is typically used to locate new oil or gas wells in water depths. It is usually employed for technical purposes. Traditional forms of such drillships were built with a modified tanker hull that had been changed, but today they are made from scratch. Most drill–ships have a dynamic positioning system that helps them stay in place over the well. They can drill up to 3,700 m down into the water.

6) Floating production systems

FPSOs are the most common types of floating production systems. FPSOs are large structures with a single hull that are usually shaped like ships and have processing facilities. These platforms stay in one place for long periods of time and do not drill for oil or gas. These systems are utilized for storing purpose and requires few components. This is one of the best places to get production that floats.

7) Tension-leg platform

TLPs are floating platforms that are tethered to the seabed in a way that keeps the structure from moving up and down. TLPs are used in water that is about 2,000 meters deep or less. The "normal" TLP is made up of four columns and looks like a semi–submersible. The Sea star and MOSES mini TLPs are proprietary versions. They are relatively cheap and are used in water depths between 180 and 1,300 meters. Mini TLPs can also be used as platforms for utilities, satellites, or early production for larger discoveries in deep water.

8) Spar platforms

Spars are moored to the seabed like TLPs, but where a TLP has vertical tension tethers, a spar has more traditional mooring lines. Spars have been made in three different ways so far: the "conventional" one-piece cylindrical hull, the "truss spar," in which the middle section is made up of truss elements connecting the upper buoyant hull (called a hard tank) with the bottom soft tank containing permanent ballast, and the "cell spar," which is made up of multiple vertical cylinders. The spar is more stable than a TLP because it has a large counterweight at the bottom and does not rely on the mooring to keep it standing. It can also move horizontally and place itself over wells far from the main platform by adjusting the tension of the mooring lines using chain-jacks attached to the mooring lines.

(for more information about oil exploration offshore platforms, please scan

4.3 Underground and under sea oil pipe lines

4.3.1 How are subsea gas pipelines built

Sea can go several kilometres deep. Putting a pipe down at the bottom is hard. On the bottom of the North Sea, however, there are 6,000 kilometers of pipelines, some of which have been there for 40 years. Solitaire is the biggest ship in the world. It is 300 meters long and 40 meters wide. The Nord Stream gas pipeline is being built with the help of this ship. The working of oil pipe lines under the sea is portrayed in Fig. 4.4.

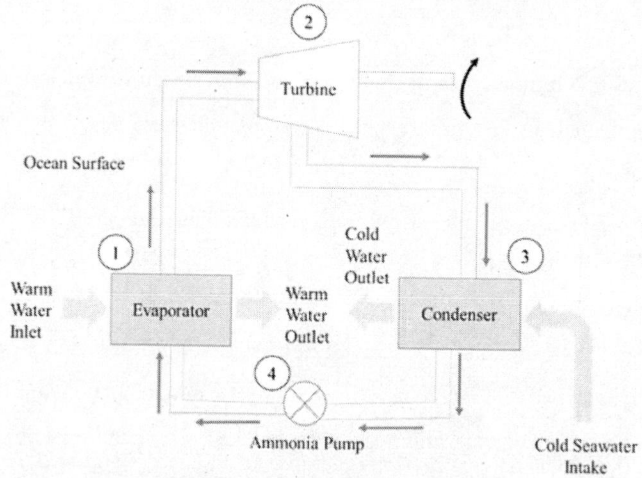

Fig. 4.4 structure of under–sea oil pipe–lines

4.3.2 Searching for obstacles

Subsea gas pipelines are responsible for 45 percent of the natural gas that is sent to Europe. Before putting the gas pipeline down, the seabed is carefully surveyed along the whole route. Specialists must find all possible problems, such as sunk ships, ammunition, or just big rocks. These problems are either taken care of or worked around. At this point, experts also figure out where the pipelines should be buried or covered up. All pipes that will be used in future gas pipelines have to go through a special process. On the inside, they have a coating that prevents friction, which makes them easier for gas to move through. Pipes are covered on the outside with a coating that prevents corrosion and then with a coating that makes the pipe sink.

1) Floating houses

Pipes are laid on the seabed by ships that are made for that purpose. Pipelaying vessels are big, floating platforms that can hold a lot of people at once. Most of the time, pipe laying involves more than one ship. Special barges keep sending pipes to a ship that is laying pipes. A ship that is watching the seabed comes before the pipe-laying ship. The pipes are unloaded directly onto the deck of the ship that is laying the pipes. They should have enough pipes for a 12-hour shift.

2) How pipes are laid

A special conveyor is built into the pipelaying vessel, which is also where the pipes are welded. Then, each weld is checked for flaws using ultrasonic waves. After welding, an anticorrosion coating is put on all the joints. Over the conveyor, welded pipes are moving towards the back. Here is Stinger, a special boom that is at an angle in the water and over which the pipes are slowly lowered to the seabed. It says how much the top part of the gas pipeline needs to bend, which keeps the metal from getting bent out of shape. Pipes usually rest on the seabed because they are heavy enough to do so. They do not need to be fixed because, after the concrete coating is added, each pipe weighs several tonnes. Pipes are laid in special trenches and backfilled with the ground only in some locations, e.g., at landfalls, to ensure the stability of the pipe.

3) From sea to shore

Most of the time, putting in a subsea gas pipeline does not start on land, as you might think, but in the water. A gas pipeline can be made up of several sections that are built from different ships at different times and then joined together. For different parts of a gas pipeline to be able to handle different pressures, pipes with different wall thicknesses are used.

When the work on the subsea section is done, pipes are slowly brought to shore with the help of a special winch that is set up on solid ground and connected to the pipe with steel ropes. The pipeline is then tied in with the part of it that is on land.

The gas pipeline must be tested for water leaks. To do this, the pipeline is filled with water under the right amount of pressure and left for a while to look for any problems. After the gas pipeline is put into service, its condition is also carefully watched. In this process, special electronic tools are used to check the pipes from the inside.

4.4 Solar chimney and sustainable architecture

A solar chimney, also known as a thermal chimney, is a way to enhance air circulation in structures by convection of air warmed by sun's electricity. A solar chimney is simply a longitudinal shaft that utilizes the sun's rays to improve natural stack air circulation through a tower. The solar chimney has been utilized for centuries, most notably by the Persians in Asia, and by the Romans in Europe.

In its most basic form, the solar chimney is a black-painted chimney. Sunlight heats up the chimney and the air inside it throughout the day, causing an airstream of air in the chimney. The vacuum generated at the base of the chimney can be utilized to oxygenate and cool the structure below. In most regions of the world, harnessing wind energy for air circulation, such as with a windcatcher, is simpler, but on warm low wind times, a solar chimney can provide airflow where there would otherwise be none.

There are, however, several types of solar chimneys. A solar chimney's basic design components are as follows:

The solar collector area: This can be situated at the top of the chimney or throughout the shaft. This element's alignment, ceramic coating type, insulation, and thermal characteristics are critical for harnessing, preserving, and leveraging solar heat gain.

The main ventilation shaft: This building's location, length, cross–sectional area, and thermal characteristics are also critical.

The inlet and outlet air apertures: The dimension, placement as well as aerodynamic prospects of such components are crucial.

Solar chimneys, also known as heat chimneys or heat stacks, can be employed in architectural settings to reduce the amount of energy consumed by mechanical systems. Air–conditioning systems and mechanical ventilation have been the classic method of climate control in several types of buildings, particularly offices, in developed countries for years. Polluted air and the redistribution of power generation have led to a fresh environmental method for building design. Technological innovations are usually coupled with bio–climatic fundamentals and common design methodologies to develop new and possibly effective conceptual designs. The solar chimney is one of these ideas that researchers and developers are actively researching, primarily through experiments and research. The process of heat exchange in a solar chimney is shown in Fig. 4.5.

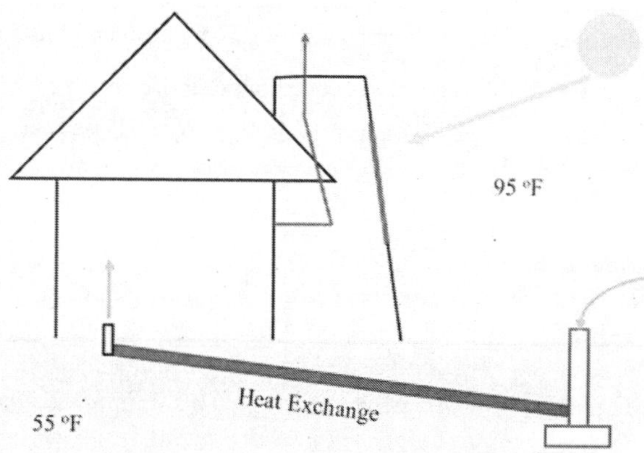

Fig. 4.5 Schematic of the solar chimney

A solar chimney can be used for a wide range of uses. Direct sunlight warms the gases within the chimney, causing it to ascend and draw air in from the bottom. This air drawing can be utilized to oxygenate a house or office, to draw air through a geothermal heat exchange, or to vent only one area, such as a composting toilet.

Fresh air can be achieved by installing vents in a building's upper level that let hot air circulate by convective heat transfer and fleeing to the outside world. In the meantime, time, cold air can be pulled in through a lower-level ventilation system. Plants on that portion of the structure could be cultivated to provide shadows for the cooler outside air.

A solar chimney can help to supplement the naturally ventilated procedure. The chimney must be significantly high than the upper levels and built on a side facing the direction of the sun's rays. Heat absorption from solar radiation can be enhanced by placing a coated surface on the sun-facing side. On the opposite side, a heat-absorbing substance can be utilized. The diameter of the chimney is less essential than the dimensions of the heat-absorbing surface. A large area of the surface enables more effective energy transfer with the air, which is required for solar radiation heating. The heat from within the chimney will increase convection and, thus, air circulation through the chimney.

(for more information about a solar chimney, please scan):

4.5 Wave Energy Caisson

The wave energy caisson is made up of a bottom box that measures 23.2m x 17m x 3m high and supports a 12m high Compartment with a curtain wall in the front and guide walls on both sides to allow waves to enter. A dual cubic curve concrete drone 10m x 7.75m at the bottom, lessening to a 2.0 m diameter circle at the top and 3m high, supports the power module. The caisson top is +5.00m above the still water level.

4.5.1 Design Parameters

1) Operating condition:

The structure is predicted to produce a peak power of 150 kW at a wave height of 1.52 m and a design wave period of 10 seconds.

2) Design extreme conditions:

The structure non-breaking and breaking wave heights were approximated to be 7.0 m. with a period of 10 s predicated on wave data collected off the coast of Trivandrum.

4.5.2 Estimation of wave forces

It is difficult to estimate wave forces on large rectangular or square caissons. Unlike in the case of cylindrical cylinders, the incident wave orientation has a substantial impact on the forces. Because the OWC caisson has an opening on one side, estimating wave forces becomes difficult due to the complex fluid flow and wave oscillations inside.

All recognized processes for estimating non-breaking wave force were tried, and the proposed method by Isaacson was finally utilized. The maximum wave force equals 14000 kN. Eventually, the framework was created to withstand 15000 kN of force.

The front lip wall is an important component of the caisson. As per Minikin, the average total dynamic pressure at SWL due to a breaking wave amplitude of 7m is 1.18 MPa, and the lip has been planned for this. The overall breaking wave force is approximately 30000 kN. Extensive metrics from an experimental framework and large model research in the research lab are required for future models to evaluate breaking wave forces.

4.5.3 Scour Protection

Scour Protection modeling at the Ocean Engineering Centre demonstrates that scour is a linear function of current velocity, with the greatest scour occurring at points 45 degrees to the direction of flow. The combination of waves on current causes a 20% to 62% rise in scour depth. The currents were discovered to be extremely low at the caisson place.

4.6 Coastal Installations for Tidal Power

By employing a variety of techniques, tidal power or tidal energy is converted into usable types of power.

Tidal energy can provide power in the future even if it is not now regularly utilized. Compared to the wind and the sun, tides are easier to forecast. Tidal energy is one of the renewable energy sources that has historically been bound in accessibility resulting in high cost and lack of places with high enough tidal amplitude or flow velocities. Nevertheless, numerous latest developments and enhancements, both in architecture (such as dynamic tidal power and tidal lagoons) and turbine engineering (such as new axial turbines and cross-flow turbines), suggest that the total availability of tidal power may be much higher than previously supposed and that financial and environmental expenses may be reduced to comparable levels.

In history, tidal mills were employed on the Atlantic coast of North America and in Europe. When the tide runs out, it rotates waterwheels that employ mechanical energy to grind grain. The incoming water was held in enormous holding ponds. The oldest examples can be found in medieval literature or even in Roman literature. In the 19th century, electricity was first produced in the United States and Europe using falling water and rotating turbines.

Tide mills have traditionally been used on both the European and North American Atlantic coasts. The onrushing water was held in large capacity pools of water, and as the tide went out, it turned waterwheels that milled grain using mechanical energy. The oldest happenings can be traced back to the medieval period or even to the Roman era. In the nineteenth century, falling water and twisting wind turbine were used to generate energy in the United States and Europe.

In 2018 and 2019, maritime technologies are expected to provide an additional 13% and 16% of the world's electricity, respectively. To further reduce costs and advance on a wide scale, R&D-promoting policies are required. France's Ranke Tidal Power Station, which started operating in 1966, was the first significant tidal power facility in the world. Up until Sihwa Lake Tidal Power Station in South Korea opened in August 2011, it was the greatest tidal power station in terms of output. Sea wall defense obstacles with ten turbines producing 254 MW are used at the Sihwa station.

Oceanic tides on Earth are used to generate tidal energy. The periodic fluctuations in the gravitational attraction that celestial bodies exert lead to tidal forces. In response, the oceans of the planet experience comparable movements or currents. As the Earth spins, this causes regular fluctuations in sea levels. Due to the predictable rhythm of the Earth's rotation and the Moon's orbit around the Earth, these changes occur with a high degree of regularity and predictability. This motion's size and variability are caused by the shifting relationships between the Moon and Sun and the Earth, the impacts of the planet's rotation, and the regional topography of the seabed and coasts.

The energy inherent in the orbital parameters of the Earth-Moon system, and to a lesser amount in the Earth-Sun system, is exclusively used by tidal power as a source of energy. Solar, wind, biofuel, wave, and other natural energies that are used in human technology all derive directly or indirectly from the sun, as do fossil fuels and traditional hydroelectric power. While geothermal energy makes use of the Earth's interior heat, which is a combination of leftover heat from planetary accretion (approximately 20%) and heat created via radioactive decay (80%), nuclear energy uses the Earth's mineral resources of fissionable components.

Electricity is produced using a tidal generator by converting the energy of tides. The potential of a location for tidal energy generation can be greatly increased by more tidal variation and higher tidal current velocities.

Tidal power is regarded as a renewable energy source since it is almost limitless and is ultimately caused by gravitational interaction with the Moon, Sun, and Earth's rotation. The Earth-Moon system loses mechanical energy as a result of tidal movement because of the pumping of water over physical barriers along coasts and the subsequent viscous dissipation at the bottom and in turbulence. In the 4.5 billion years since the Earth's creation, this energy loss has led to a slowing of the planet's rotation. The Earth's period of rotation (the length of a day) has risen from 21.9 hours to 24 hours over the past 620 million years; during this time, the

Earth-Moon system has lost 17% of its rotational energy. Despite tidal power will consume more energy from the system, the impact will be insignificant and will not be evident soon.

(for more information about tidal power, please scan):

4.7 Wind towers

The windmill is a device that utilizes sails or blades called vanes to convert wind energy into rotational energy in order to grind grain (gristmills), while the name is also used to refer to wind pumps, wind turbines, and other applications. Such machines are commonly regarded as wind engines.

The horizontal or panemone windmill initially arose in Greater Iran in the ninth century, while the vertical windmill was first presented in northwest Europe in the 12th century. Windmills were utilized throughout the high medieval and early modern periods. Nowadays in the Netherlands, there are about 1,000 windmills, which are acknowledged as symbols of Dutch culture.

4.7.1 Horizontal windmills

The earliest practicable windmills were panemone windmills, which used sails that revolved in a horizontal plane around a vertical axis. Such windmills were employed to crush grain or draw up water. They were constructed of six to twelve sails wrapped with reed matting or textile material. Based on the caliph's talk with a Persian slave builder, a medieval story claims that windmill technology was employed in the Arab region under the rule of Umar ibn al-Khattab. The fact that the story about the caliph Umar was only documented in the tenth century casts doubt on its veracity. During the ninth century, the Persian geographer Estakhri wrote that windmills were employed in Khorasan. Similar windmills were widely used in the Middle East and Central Asia before being exported from those regions to Europe, China, and India. The vertical-axle windmill began to spread to Southern Europe by the 11th century, covering the Iberian Peninsula and the Aegean Sea. Following Yelü Chucai's visit to Turkestan in 1219, a horizontal windmill with rectangular blades of a similar design that was used for agricultural purposes was also discovered in the thirteenth–century in China. The layout of a horizontal windmill is shown in Fig. 4.6.

In Europe during the 18th and 19th centuries, vertical-axle windmills were erected in modest numbers, such as Fowler's Mill in Battersea, London, and Hooper's Mill in Margate, Kent. The above first recent systems do not appear to have been independently created by 18th-century engineers, but rather to have been directly influenced by the vertical-axle windmills of the medieval era.

In Europe during the nineteenth and twentieth centuries, vertical-axle wind turbines were constructed in tiny groups, such as Fowler's Mill in Battersea, London. Such early medieval case studies appear to have been independently invented by 18th-century engineers rather than being certainly impacted by the medieval vertical-axle windmills.

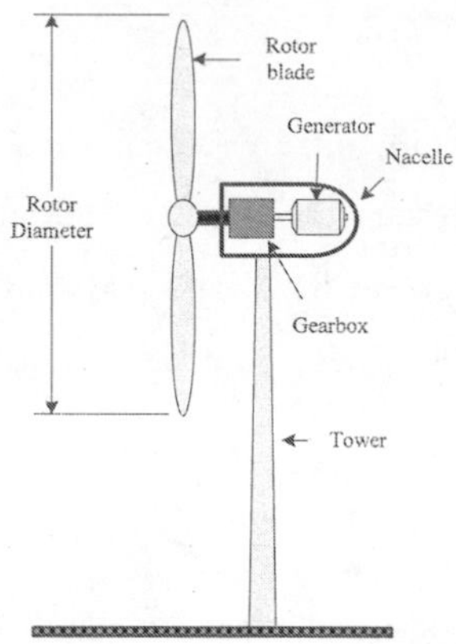

Fig. 4.6. Layout of the horizontal windmill

4.7.2 Vertical windmills

The vertical windmill is a 12th-century invention that was initially utilized in the triangle formed by northern France, eastern England, and Flanders. Its name comes from the plane on which its sails rotate. It is unknown if the horizontal windmill's introduction to Southern Europe in the century before had any bearing on the development of the vertical windmill.

In Yorkshire's historic village of Weedley, which was perched on the southernmost point of the Wold and overlooked the Humber Estuary, the oldest known mention of a windmill (assumed to be of the vertical variety) dates to 1185. There have also been a variety of early, less precisely documented 12th-century European references that mention windmills. Cereals were crushed at these first mills. The layout of vertical windmills is shown in Fig. 4.7.

(for more information about windmill towers, please scan):

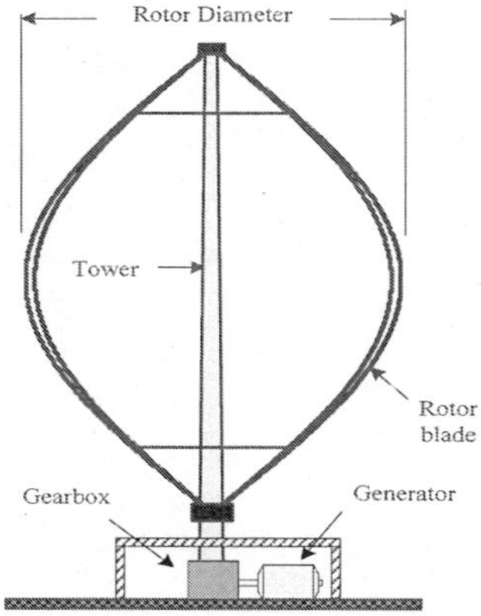

Fig. 4.7. Layout of the vertical windmill

4.8 Underground and aboveground hydro power stations

4.8.1 Underground hydro power stations

An underground power station is a form of hydroelectric power station that is built by digging the main parts from a rock as opposed to the more usual surface-based building techniques (such as the machine hall, penstocks, and tailrace).

The decision to build a power station underground is influenced by one or more considerations. A surface power plant may not be feasible in gorges or steep valleys, and thus, the topography or geology surrounding a dam is properly considered. It could be cheaper expensive to build a power plant underground than on loose soil at the surface. A surface station is frequently impossible in valleys that are vulnerable to avalanches. Large hydroelectric power plants were more frequently buried after World War II to shield them from bombings. One of the underground hydro power stations is shown in Fig. 4. 8.

Some notable underground power stations are:

One example of an underground pumped storage plant is Kazunogawa Power Station in Japan which generates 1,200 MW of electricity. It is made up of four 400 MW power plants. The underground station's chamber is 500 m below ground. It measures 210 m elongated, 54 m in height, and 34 m in width. The height of the head is 714 m.

The second largest underground power station is in Newfoundland and Labrador, Canada. It has 11 turbines that generate 5,428 MW. The powerhouse is 232 m long, 148 45 m tall, 19 m wide, and 330 m underground. The length of the two tailrace tunnels is 1691.64 m. The net head measures 312.42 m.

Fig. 4.8. Underground hydro power station

4.8.2 Aboveground hydro power station

Electricity produced from hydropower is known as hydroelectricity or hydroelectric power (water power). Nearly 4500 TWh of electricity will be generated by hydropower in 2020, which will be greater than nuclear power and all other sources of renewable energy included. This represents one–sixth of the world's electricity production.

One of the most important components of developing safe and reliable electrical supply systems is hydropower, which can deliver significant volumes of low-carbon electricity on demand. Since the quantity of energy generated may be raised or decreased in response to changing electricity demand, a hydroelectric power plant with a dam and reservoir qualifies as a flexible source. Once built, a hydroelectric complex does not generate any trash directly and nearly always generates less greenhouse gas than electricity facilities that use fossil fuels. However, significant volumes of greenhouse gases may be released if buildings are built in lowland rainforest regions where part of the forest is submerged.

The loss of agricultural land and population displacement are the two main environmental effects of the construction of a hydropower complex. Furthermore, they disrupt the river's natural ecology by exacerbating silting and sedimentation styles, as well as changing ecosystems and biodiversity. While dams can decrease stormwater runoff, they also pose the possibility of a major collapse. The biggest power-generating facility of any sort in the world is the Three Gorges Dam in Central China. One of the aboveground hydro power stations is shown in Fig. 4.9.

(for more information about hydroelectric power plants, please scan):

Fig. 4.9. Aboveground hydro power station

4.9 Dams, Tunnels and Penstocks

4.9.1 Dams

A dam is a natural wall that halts or inhibits the movement of groundwater. Dams create storage tanks that can be used for a variety of purposes, which would include flood protection, agriculture, public consumption, manufacturing use, fish farming, and navigation. Dams and hydroelectric power are regularly used in conjunction to make power. A dam can collect or store water that can be evenly distributed between regions. While some other frameworks, such as dikes, are utilized to regulate or prevent river flow into geographical regions. Dams are frequently employed to store water. The world's oldest dam is the Jordanian Jawa Dam, which goes back to 3,000 BC. The term dam may be found in the names of numerous historic cities, including Amsterdam and Rotterdam, and can be traced back to Middle English and Middle Dutch earlier.

Kinds of dams

Dams can be formed by individual interaction, natural occurrences, or by the participation of animals. Typically, human-made dams can be categorized based on their height, intended use, or structural design.

Dams are categorized as conveniently constructed without materials, arch-gravity, embankment, or masonry dams, with various subtypes, depending on the construction and material utilized.

4.9.2 Tunnels

A tunnel is a corridor that has been carved through the soil, rock, or other surrounding material. It is enclosed, save for the entrance and exit, which are often located at either end. Even while some modern tunnels were built using immersed tube construction techniques rather than conventional tunnel boring techniques, a pipeline is not a tunnel.

A tunnel may be used for canal flow, rail traffic, or foot or vehicle traffic on the road. Typically, the tunnel is where the core of a rapid transit system is located. Some tunnels are utilized as aqueducts or sewers to deliver drinking water or to power hydroelectric plants. Utility tunnels are employed to join buildings for easy movement of people and components, as well as to route steam, chilled water, electrical power, or communications cables.

Secret tunnels are constructed by civilians or the military for the transportation of persons, contraband, or weapons. To securely allow wildlife to traverse obstacles established by humans, special tunnels like wildlife crossings are constructed. Tunnel networks are structures that link tunnels. The recently built tunnel at Pragati Maidan, New Delhi, is shown in Fig. 4.10.

Fig. 4.10. Tunnel at Pragati Maidan, India

4.9.3 Penstocks

A penstock is a sump pit, gate, intake configuration, or sealed transit point that controls water flow and provides water to hydropower plants and drainage systems. The word is a holdover from previous watermill and mill pond technologies.

Penstocks for hydroelectric plants often have a surge tank and a gate mechanism. Depending on the application, they may consist of a variety of parts, including anchor blocks, drain valves, air bleed valves, and support piers. Flow is controlled by the functioning of the turbines, and it is zero when the turbines are not in use. Because penstocks are used in dirty water systems, they must be preserved through heated water washing, manual washing, wear–resistant coatings, and evaporation.

The phrase is also employed to describe the pathways that lead to and from high-pressure sluice gates in agricultural dams.

Penstocks are also employed in the building of mining tailings dams. The penstock is typically constructed utilizing penstock rings and placed very near the sediment dam's center. These regulate the volume of water, allowing the slimes to float to the surface. A penstock pipeline is then used to transport this water back to the plant under the tailings dam.

In order to manage the water flow through the mill wheel, penstocks are frequently employed at mill sites.

In water management systems like surface water drainage and unclean water sewers, penstocks are frequently employed. Penstocks offer a way to isolate flows and control the flow of water as it is sent to landfills or power plants

Summary

In this unit, technologies and associated tools used to develop infrastructure to build different types of power plants are discussed. Civil construction required to related with coal mining, oil exploration, offshore platforms, underground and under-sea oil pipelines, solar chimney, coastal installations for tidal power plants, wind towers and infrastructure for building hydropower plants are discussed in brief and limited to the electrical engineering point of view.

Short and Long Answer Type Questions:

Q1. Discuss different coal mining methods and tools required during the process of mining.

Q2. What may be the common obstacles in laying down the oil or gas pipe line in the seabed?

Q3. Discuss technologies and tools used to lay down the subsea gas pipelines?

Q4. What is a solar chimney? With the help of a neat sketch, discuss the different components of a solar chimney.

Q5. Discuss wave energy caisson and its design parameter.

Q6. Discuss the merits and demerits of the design of horizontal–axis and vertical–axis wind turbines.

References and suggested readings

1) R. D. Singh, “Principles and Practices of Modern Coal Mining,” New Age International (P) Ltd.
2) Rahman M.M. “Cold Inflow Free Solar Chimney Design and Applications,” Springer.

3) Robert Gasch, Jochen Twele, “Wind Power Plants: Fundamentals, Design, Construction and Operation,” Springer.
4) Bikash Pandey, Ajoy Karki, “Hydroelectric Energy, Renewable Energy and the Environment,” CRC Press.

****

5 Building Structure for Nuclear Power Plants

UNIT SPECIFICS

In this unit, the following topics have been discussed related to building structures for nuclear power plants:

- Structural components of a nuclear power plant.
- Design and construction constraints of the reactor.
- Testing of reactor containment buildings
- Requirements in the ultimate limit state, deformations, and vibrations.
- Safety aspects for the storage of nuclear fuel
- Waste disposal in a nuclear power plant

RATIONALE

This unit introduces the infrastructure and related civil work required to build a nuclear power plant. Due to the use of radioactive material as a fuel in nuclear power plants, designing and construction of a nuclear reactor are full of challenges to make the use of the plant safe under normal conditions as well as under natural and war tragedies. Designing, construction constraints of the reactor, testing of reactor containment buildings, safety aspects for the storage of nuclear fuel, and waste disposal in a nuclear power plant are discussed in this unit in detail.

PRE-REQUISITES

Basic Knowledge of Nuclear power plants & infrastructure used

UNIT OUTCOMES

List of outcomes of this unit is as follows

U5-O1: To know about the infrastructure & technologies used in nuclear power plants

U5-O2: To know about the designing and construction of nuclear reactors.

U5-O3: To be familiar with the testing of reactor containment buildings.

U5-O4: To know about the safety aspects of the storage of nuclear fuel.

U5-O5: To understand waste disposal in a nuclear power plant.

Unit outcomes	EXPECTED MAPPING WITH COURSE OUTCOMES (1- Weak Correlation; 2- Medium correlation; 3- Strong Correlation)					
	CO-1	CO-2	CO-3	CO-4	CO-5	CO-6
U5-O1	-	1	1	-	3	-
U5-O2	-	-	-	-	3	-
U5-O3	-	-	-	-	3	-
U5-O4	-	1	1	-	3	-
U5-O5	-	-	-	-	3	-

5.0 Introduction

A nuclear power plant is a thermal power station in which a nuclear reactor acts as its main energy source. A nuclear reactor is a device that is used to start a chain reaction. In a thermal power station, the heat produced by a chain mechanism generates steam, which pushes a steam turbine linked to an electric generator, which holds the responsibility of generating electricity.

5.1 Structure of a Nuclear Power Plant (NPP) and its components

The heat stored is transformed into mechanical energy by a steam turbine. In a nuclear power plant, the generator transforms the kinetic energy delivered by the turbine into electricity. The cooling method eliminates heat from the fuel rods and transmits it to a different area of the plant where it can be used to generate power or perform other valuable work.

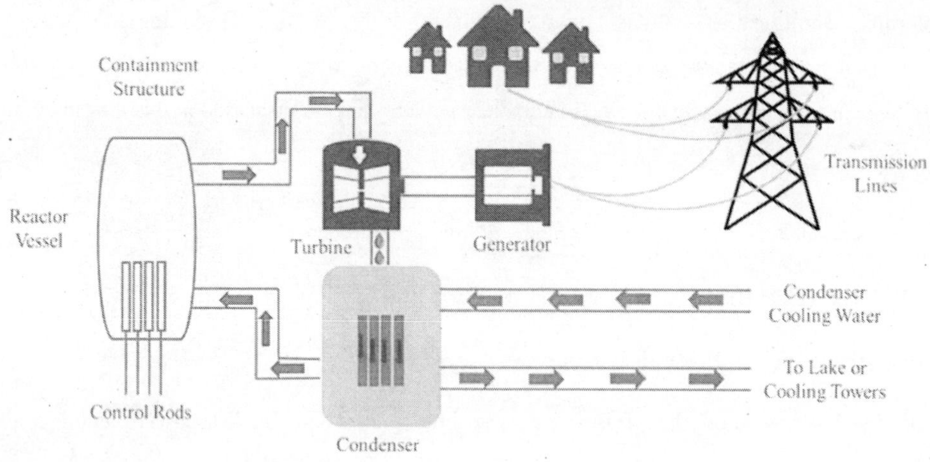

Fig. 5.1. Structure of Nuclear Power Plant

Safety valves can be used in case of emergencies to protect pipelines from ruptures or the reactor from bursting. The feed water mechanism regulates the level of water in the steam generator and nuclear reactor. Most nuclear power plants necessitate two different off-site power sources to support transformers that are kept separate in the station's switchyard and can receive power from numerous power lines.

Uranium is employed as fuel in nuclear reactors to initiate the chain of reactions. Uranium is a metallic element that is predominantly observed in seawater and rocks. The working of nuclear power plants is discussed in Fig. 5.1.

The important component of a nuclear power plant includes:

1) Nuclear Reactor: A nuclear reactor is a critical component of nuclear power stations. The primary function of a nuclear reactor is to start and regulate a long-term nuclear chain reaction.
2) Steam Generators: Steam generators transform excess heat in a nuclear reactor core into feedwater. They are placed between the primary and secondary coolant loops in pressurized water reactors.
3) Pressurizer: A pressurizer is a distinct vessel attached to the main circuit and partly full of water heated to the saturation temperature for the required pressure by underwater electrical heating systems thereby maintaining pressure in the main circuit. The temperature inside the pressurizer is kept at 345 °C which results in a subcooling margin of 30 °C.
4) Reactor Coolant Pumps: Primary coolant pumps are utilized to circulate primary coolant around the primary circuit. These pumps are strong, consuming up to 6 MW each, and can be used to heat the main coolant prior to the reactor starting.
5) Safety System: As per the policy framed by United States Nuclear Regulatory Commission, the primary goals of nuclear reactor safety mechanisms are to close the reactor, keep it closed, and protect radioactive particles from being released. Reactor safety mechanisms are made up of the following components:
    1) Reactor Protection System
    2) Essential service water system
    3) Emergency core cooling systems
    4) Emergency power systems
    5) Containment systems
6) Turbine: A steam turbine is a device that utilizes heat energy extracted from superheated

steam to perform mechanical work.

7) Generator: A generator is a device that converts mechanical energy to electrical energy.
8) Condenser: A condenser is a heat exchanger that compresses steam from the turbine's final stage.
9) Condensate-Feedwater System: Condensate-Feedwater Systems serve two primary purposes. To supply sufficient high-quality water to the steam generator and to boil the water to near-saturation temperatures.

(for more information about components of nuclear reactors, please scan):

5.2 Requirements in the Designing of the Reactor Containment

5.2.1 Containment structure design basis

The reactor containment framework, such as connection holes, cavities, and the containment heat extraction structure, must be developed so that the containment framework and its internal containers can cater to the determined temperature and pressure conditions resulting from postulated accidents without surpassing the design discharge percentage and with sufficient margin. This margin must account for (1) the effects of potential energy sources that were not considered when determining peak conditions, (2) the limited experience and experimental data available for defining accident phenomena and containment responses, and (3) the conformism of the computational framework and input variables.

5.2.2 Fracture prevention of containment boundary

The outer boundary of the reactor containment must be intended with enough margin to ensure that (1) its components work in a nonbrittle fashion under functioning, maintenance, experimenting, and hypothesized accident circumstances, and (2) the possibility of quickly proliferating rupture is reduced. The design must consider the temperatures and other requirements of the containment border substances during the procedure, conservation, testing, and presumptive accident conditions, as well as the instabilities in deciding (1) properties of materials, (2) lingering, stable, and transitory stresses, and (3) flaw dimensions.

5.3 Requirements in the ultimate limit state, deformations, and vibrations

A structure may become inadequate for usage not only when its breakdowns but also when it interrupts the usability necessities of rebounds, vibrations, cracks due to weakness, erosion, and fire.

The structure of the steel should be made in such a way that it must fulfill basic requirements such as stability, firmness, strength, usability, brittle fracture, fatigue, fire, and robustness. The structures must encounter the subsequent necessities:

a) Remain fit with satisfactorily reliability and capable of withstanding all kinds of loads.

b) Have acceptable resilience under regular maintenance.

c) Do not endure total damage or implode disparately as a result of unintended occurrences such as eruptions, vehicle impact, or the outcomes of human discrepancy to the level that it exceeds local damage.

5.4 Testing of reactor containment buildings

A containment is a structure made of armoured steel and cement that surrounds a nuclear reactor. It is intended to prevent the escape of radioactive particles or gas to a maximum pressure of 275 to 550 kPa in the event of an emergency. The final and fourth hindrance to radioactive leak follows the fuel ceramic itself, the metal fuel cladding tubes, and the reactor vessel and coolant system.

Each nuclear power plant in the United States is built to withstand specific conditions, which are referred to as "Design Basis Accidents" in the Final Safety Analysis Report (FSAR). The FSAR is typically visible to the audience at a city library near the nuclear plant.

The containment structure is typically an impenetrable steel frame that encloses the reactor, which is usually encased from the external environment. The steel can either stand alone or be attached to the concrete missile shield. In the United States, federal regulations govern the construction and width of the containment and missile shield, which must be sufficient to withstand the effect of a fully loaded airliner without bursting.

While containment is essential for the most serious nuclear reactor fatalities, it is only intended to hold or condense steam in a brief time, with long-term removal of heat supplied by other processes. The containment pressure boundary was maintained in the Three Mile Island accident, but due to insufficient cooling, radioactive gas was deliberately released out of the structure by workers to avert high pressurization. During the accident, close to 13 million curies of radiation were emitted into the surroundings.

5.4.1 Types

Nuclear power reactor containment systems can be distinguished by their dimensions, structure, metals employed, and fire suppression system. The type of structure used is defined by the reactor types, reactor production, and particular plant requirements.

Suppression systems are essential to provide a shield to the system and have a large impact on containment shape. Suppression is the process of condensing steam after it has been removed from the cooling system due to a major interruption. Because deterioration warm air does not dissipate easily, a long-term process of suppression is required, which could simply be heat transfer with the ambient environment on the wall of the containment. Based on safety purposes, containments are classified as "large-dry," "sub-atmospheric," or "ice-condenser."

1) Pressurized water reactors (PWR)

The containment of a PWR includes the steam generators and the pressurizer, as well as the whole reactor building. The missile shield that surrounds it is usually a tall cylindrical or pyramidal structure. Because the containment strategy during the leakage design basis accident encompasses getting enough for the steam/air mixture that ultimately resulted from a loss-of-coolant-accident to expand into, limiting the ultimate pressure reached in the containment building, PWR containments are usually big.

Initial prototypes by multinational companies featured a can-type structure which is made of concrete material. Because concrete has a compression ratio strength than the tensile strength, this is a conceptual model for construction materials because the exceptionally heavy upper section of the confinement imposes a huge pressure gradient that hinders some tensile stress if the confinement pressure suddenly rises. Numerous spherical-shaped containment models for PWRs have been built as reactor designs. This appears to be the most conceptual model, based on the material used, because a sphere is the best foundation for merely comprising a large pressure. Many modern PWR models combine the two, with a cylindrical bottom section and a half-spherical top. The schematic of pressurized water reactors is shown in Fig. 5.2.

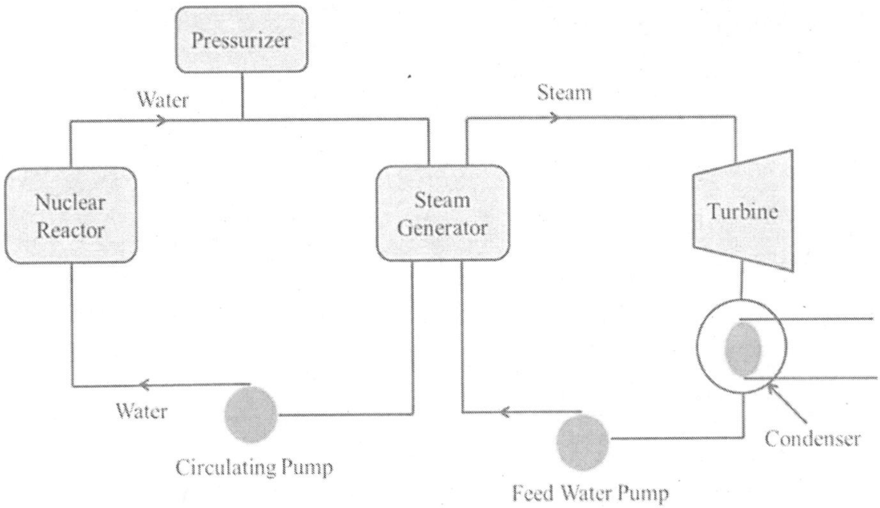

Fig. 5.2. Working of Pressurized Water Reactor

2) Light water graphite reactors (LWGR)

LWGR are created by USSR only. Secondary containment-like constructions were used in RBMK designs. The upper plate of the reactor is one of the parts of the defensive structure. During the Chernobyl disaster, the plate experienced pressures that exceeded the expected bounds and raised up. The functioning of light water graphite reactors can be understood with the help of Fig. 5.3.

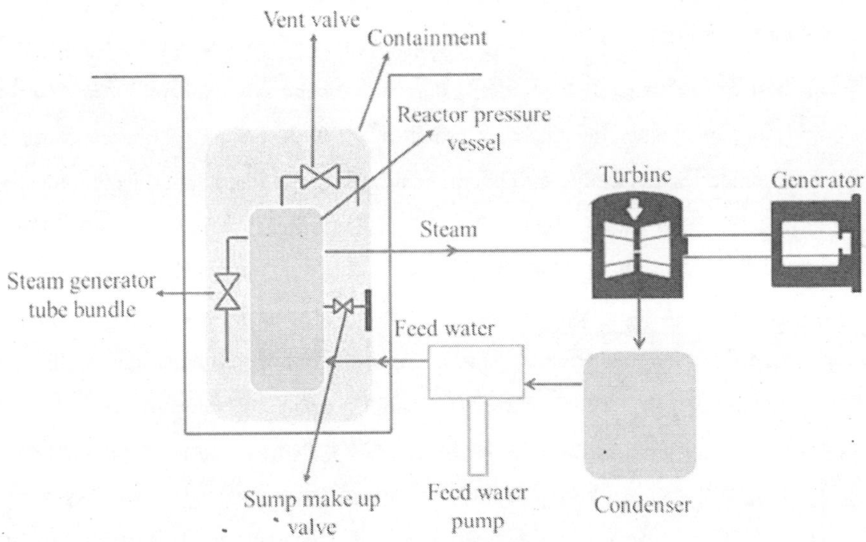

Fig. 5.3. Working of light water graphite reactors

3) Boiling water reactors (BWR)

The containment strategy in a BWR is slightly different. The confinement of a BWR is comprised of a drywell (DW), which houses the reactor, related freezing devices, and a wet well. The DW is significantly lesser than a PWR containment in size but has utmost importance. During the conceptual discharge design limits accident, the coolant in the DW flickers to steam and hence speedily pressurizing it. Vent pipes from the DW direct the steam underneath the wet well's water level, compacting the steam and restricting the pressure eventually attained. Throughout normal operating conditions and refueling, the DW and the wet well are surrounded by a secondary containment dome and kept at a slight sub-atmospheric level. The working of boiling water reactor is displayed in Fig. 5.4.

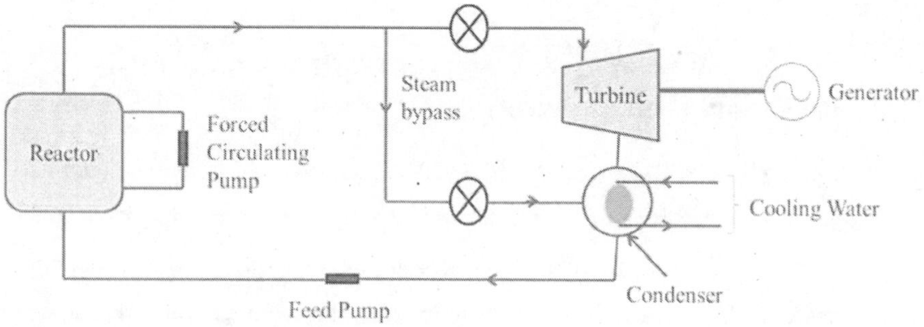

Fig. 5.4 Working of Boiling Water Reactor

5.4.2 Design and testing requirements

Title 10 of the Code of Federal Regulations, Part 50, Appendix A, General Design Criteria (GDC 54-57) or some other design basis provides the basic design requirements for the separation of boundaries perforating the containment wall in the United States. Isolation valves, configured in accordance with Appendix A, are installed on each large pipe that enters the containment, such as steam lines. One on the inside and one on the outside for smaller lines. Developers configure isolation vents where large and high-pressure lines exit containment to allow for relief valves and maintenance. These vents close quickly in the case of spills in the high-pressure piping trying to carry the reactor coolant, preventing radioactive particles from entering the reactor. Usually, vents on routes for standby systems perforating containment are sealed. Other signals, such as the containment increased pressures generated during a high-energy line break, may cause the confinement isolation valves to close. The concrete structure needs to serve to encompass the steam pressure, but such an interruption at a pressurized water reactor usually has no radiation exposure.

During normal operating conditions, the confinement is tightly sealed, only with marine-style airlocks providing access. Increased ambient temperature and core radiation restrict the amount of time individuals can spend inside confinement whereas the power station is working at full power. In the occurrence of a catastrophic emergency, the containment is intended to wrap off and encompass a meltdown. Although superfluous systems have been installed to avoid a meltdown, it is assumed that one will occur, necessitating the need for a containment dome.

(for more information about nuclear reactors, please scan):

5.5 Safely aspects for the storage of nuclear fuel

The International Atomic Energy Agency (IAEA) defines nuclear safety as "the attainment of appropriate operating environment, the avoidance of fatalities, or the remediation of fatality risks, arising in the safeguarding of workers, the public, and the surroundings from inordinate radiation risks." Nuclear security is defined by the IAEA as "the mitigation, sensing, and reaction to stealing, subvert, illegal access, illegal transfer, or other harmful behaviour involving nuclear materials, other radioactive elements, or their related facilities."

This includes nuclear power plants and other nuclear facilities, nuclear material transportation, and the utilization and collection of nuclear materials for healthcare, energy, market, and military operations.

The nuclear power industry has continued to improve reactor efficiency and safety while also proposing fresh and more secure nuclear plants. However, complete security cannot be assured. Due to human error and external factors with a bigger influence than expected are possible sources of issues: the creators of the reactor designs at Fukushima in Japan did not realize that a tsunami produced by a seismic event would deactivate the backup power that was purported to stabilize the nuclear power plant after the seismic events. Terrorist attacks, battle, insider subvert, and cybersecurity threats are all possible disaster situations.

For varied purposes, such as confidentiality, nuclear weapon security, and armed services research involving nuclear materials are usually managed by authorities separate from those in charge of civilian safety. Terrorist groups procuring nuclear bomb-making substance is still a source of anxiety.

Nuclear safety procedures occur in a variety of circumstances as of 2011, such as:

- Use of fission reactions in power stations, submarines, and ships.
- The extraction, collection, and usage of fuels such as uranium-235 and plutonium-239
- Nuclear waste, the radioactive waste residue of nuclear materials
- Continuity of uranium supplies

Except for thermonuclear weapons and exploratory fusion investigations, all nuclear power safety concerns originate from the requirement to restrict physiological take-up of committed dose and external beam radiation dosages due to radioactive material.

(for more information about safety issues, please scan):

5.6 Waste disposal in a nuclear power plant

In the absence of a recycling process, spent fuel is strongly rejected and must be kept in a separate plant for complete removal. Furthermore, the waste stream generated by spent-fuel recycle should be discharged. Many nuclear countries have investigated waste treatment techniques and geomorphologic places, but no lasting management of waste is currently available that is being used anywhere on the globe. All spent nuclear fuel and packaged waste are stored in refrigerating pools or aboveground storage barrels awaiting approval and building of disposal facilities.

5.6.1 Waste conditioning

Spent fuel should be secured in canisters that are estimated to survive for hundreds of years. However, no perpetual waste disposal is fully active, the basic method for preparing spent fuel for waste is presumed to stay the same. The fuel pins are to be eliminated from their components after one to five years of aboveground storage. End manifolds and fuel assembly elements that do not include fuel will be eliminated, and the pins will be resealed into a closely packed structure and arranged in a highly corrosive stainless-steel compartment. A cover will be soldered on, and the compartment will be protected by an overpack. The steps involved in the conditioning of waste in a nuclear power plant is described in Fig. 5.5.

Each year, some effluents are is produced as a result of the fission-product solution that results from removing impurities. One glass manufacturing procedure for classifying these wastages is commercially available in France, the United Kingdom, and Japan, and has been examined in numerous other countries. The disposal solution is entirely disappeared, abandoning the fission products in a solid residue that is warmed until each of the component nitrate salts has been turned to oxides. These oxides are then placed in a glass-forming cooker and blended with ingredients to produce borosilicate glass. As the glass forms, the fission-product oxides disperse. The molten glass metal is then immersed into a steel cylinder that measures 200-400 mm in diameter and 1 meter in height, where it reinforces. The solid canister-like component is prepared for waste after being encased with an overpack of bentonite clay.

(for more information about waste conditioning, please scan):

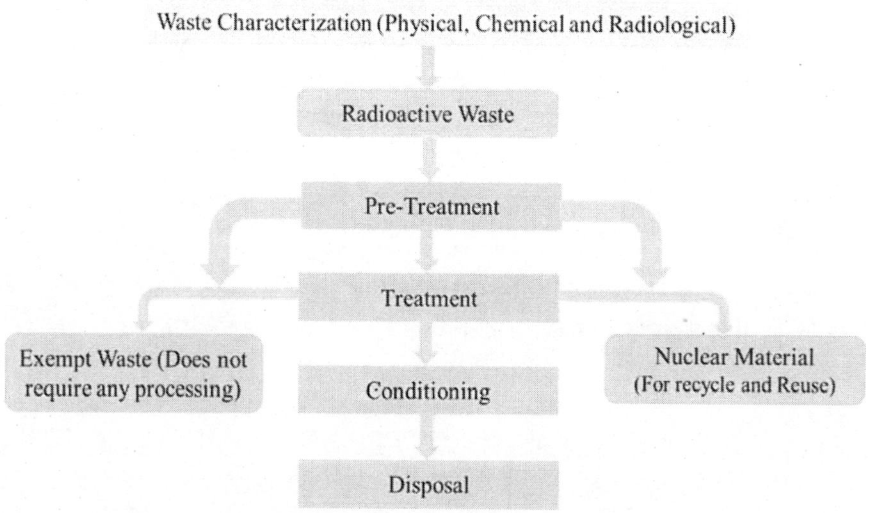

Fig. 5.5 Typical layout for waste conditioning

5.6.2 Risks of nuclear waste disposal

When viewed holistically, the dangers of nuclear waste disposal appear exceptionally low; however, among the wider populace, these dangers are one of the most dangerous attributes of the nuclear fuel cycle. Nuclear waste preserves its high level of radioactive material over several hundred years, but after a thousand years, the leftover radioactive waste is comparable to that of an equal amount of natural uranium ore. This divides the safety issue into two periods of time the first millennium, throughout which rigid preservation of the waste products in the storage site is critical, and the second millennium, throughout which it is only necessary to make sure that any discharge that happens is tiny and safe.

The perforation of underground water into the buried disposal facility, accompanied by the rust of the waste casings, releasing of waste products, and the release of the sample solutions to the surroundings, is one possible path for the onset of nuclear waste to the exterior. On the other hand, water relocates slowly in most rock formations, and, contrary to common belief, even in the most extreme scenarios, only decreased amounts of pollutants are anticipated.

5.7 Waste Disposal and Safety Aspects in Nuclear Power Plants in India

Any enterprise asset management causes the production of waste. The nuclear industry is not an exception, and the occurrence of radiation radiating nuclear material that may have a negative consequence for living creatures and are expected to be passed down to future generations is what distinguishes nuclear or radioactive wastes from other traditional toxic waste. The other distinct characteristic of nuclear waste is the degradation of radioactive material over time. This reality is profitably misused by nuclear waste management teams.

In the Indian context, radioactive waste management encompasses all kinds of radioactive waste produced throughout the nuclear fuel cycle, from uranium mining to fuel manufacturing to reactor procedures and successive spent nuclear fuel upcycling. Because spent fuel is used again in order to retrieve and reprocess the U and Pu generated, the fuel cycle is referred to as 'closed,' as opposed to other nations where spent fuel is preserved as waste. Fig. 5.6 shows all the operations in India's closed fuel cycle, as well as their interconnection. Radioactive waste is also produced as a result of the utilization of radioactive elements in healthcare, market, and scientific studies.

Before the final storage/disposal, radioactive waste products must be differentiated, categorized, managed, handled, accustomed, and supervised. Radioactive wastes can be solid, liquid, or gas, and have a wide range of physical and chemical characteristics. Radioactive wastes are divided into three categories based on their radioactivity level: exempt waste, low and intermediate level waste, and high-level waste. Fig. 5.7 illustrates the categorization of radioactive waste as advised by the IAEA.

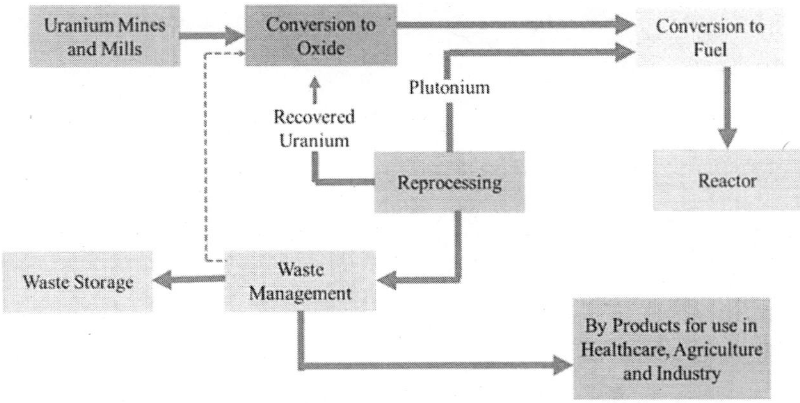

Fig. 5.6 Nuclear Fuel Cycle

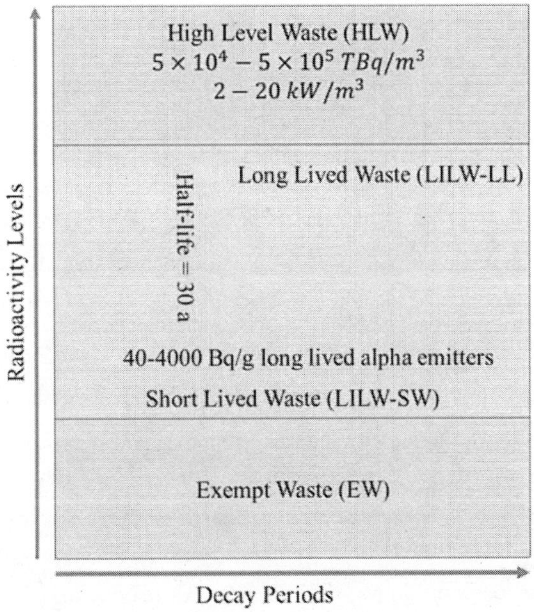

Fig. 5.7 Classification of the radioactive waste

Exempt wastes have radioactivity levels that are too low for regulators to be concerned about. These can be discharged in the surroundings without harming the environment.

Low and intermediate level wastes are further classified as either short or long-lived wastes. Radiological risk connected with short-lived wastes (30 years half-life) are drastically decreased by radioactive decay over a few 100 years. High-level waste includes both short- and long-lived radioactive particles, necessitating strict exclusion from the biodiversity and, in most cases, treatment of waste into profound geological formations. High level waste typically creates heat due to the release of a considerable amount of energy in the radiation form.

During the reuse or recycling of spent nuclear fuel, high level radioactive liquid waste (HLW) is generated, which contains most of the radioactive material in a whole fuel cycle. Furthermore, hull waste is produced as solid HLW after the spent fuel is dispersed for reuse or recycling.

Presently, the Nuclear Power Corporation of India Limited (NPCIL) oversees all NPPs in India. As a result, any conversations about NPPs in India will be about it. NPPs in India are not only secure, but are also well-regulated, with appropriate radiation safety for workers and the general public, frequent monitoring, measurements, standard specifications maintenance and operation methods, a well-defined waste management technique, and suitable, well-documented, and offer regular emergency preparedness and disaster management strategies. The NPPs have occupational health policies that include periodic health checks, measurements, and bioactivity, and they are supported by properly equipped Personnel Decontamination Centres staffed by doctors with Occupational and Industrial Health certifications. Furthermore, they have received specialized training in dealing with radiological emergencies.

In all operations, safety is given the utmost importance. All nuclear facilities are positioned, constructed, built, procured, and functioned in full compliance with stringent safety and quality requirements. All nuclear facilities and their systems/components are designed with defense in depth, failover, and diversification in mind. The country's regulatory structure is strong, with the individual Atomic Energy Regulatory Board (AERB) having the authority to structure guidelines, establish security guidelines and requirements, and supervise and enforce all safety provisions.

The AERB shall exercise state oversight through a staged licensing scheme. As a result, India has an outstanding safety record in over 277 reactor years of operation of power reactors.

The Atomic Energy Act of 1962 governs the creation of nuclear power. The Atomic Energy Act incorporates all atomic energy-related operations, which include power production.

(for more information about the safety of NPP in India, please scan):

Summary:

In this unit, construction and components of a nuclear power plant are discussed. Generating power from a radioactive material like uranium and its allies is a challenging task as it involves radiation of dangerous radioactive participles. Safety aspects in the construction of a nuclear reactor and disposal of waste produced in the process of power generation are the key challenges. The focus of this unit has been on the designing and construction of a safe nuclear power plant. Testing procedures for reactor containment buildings are also discussed.

Short and Long Answer Type Questions

1. Enumerate the basic factors to be considered in the design of nuclear reactor.
2. What are the special facilities required for a nuclear power station with respect to fabrication of fuel elements and handling of radioactive material?
3. How waste is disposed-off in a nuclear power plant? What arrangements are made on the site of a nuclear power plant for safe disposal of the waste?
4. What important considerations are required in the design of reactor containment structure's border?
5. How can nuclear power reactor containment systems be classified?
6. Write a note on safety aspects defined by International Atomic Energy Agency.
7. Discuss the testing procedures for reactor containment buildings.

References and Suggested Readings

1) Amelia Frahm, "Nuclear Power: How a Nuclear Power Plant Really Works," *Nutcracker Publishing Company.*

2) Rüdiger Meiswinkel, Julian Meyer, Jürgen Schnell, "Design and Construction of Nuclear Power Plants," *Wiley.*

3) Christine Honders, "Nuclear Power Plants: Harnessing the Power of Nuclear Energy: Harnessing the Power of Nuclear Energy," *The Rosen Publishing Group*

****

6 Engineering For Energy Conservation

(Green Building and Green Architecture)

UNIT SPECIFICS

In this unit, the following topics have been discussed for basic understating related to energy conservation in buildings from structural, architectural, and material points of view:

- Concept of green building and green architecture
- Materials for building from a green energy point of view
- Designing and Building Orientations to meet the concept of green architecture
- Leadership in Energy and Environmental Design (LEED)
- Embodied Energy Analysis and tools used

RATIONALE:

Energy conservation in buildings, especially high-rise commercial buildings, is essential in all countries. About 15 to 20 % savings in energy are possible if buildings are designed from a green architectural point of view. The concept of green energy, green architecture, and structural material required that lead to savings in energy and help in protecting the environment is described in this unit. Embodied Energy Analysis and tools used are also discussed.

PRE-REQUISITES

Basic Knowledge of Environmental Science and Energy

UNIT OUTCOMES

List of outcomes of this unit is as follows

U6-O1: To know the basic concept of energy conservation in buildings.

U6-O2: To understand the concept of green building and green architecture.

U6-O3: To be familiar with LEED ratings.

U6-O4: An idea about energy related enterprises.

U6-O5: Knowledge about embodied energy analysis and tools used.

Unit outcomes	EXPECTED MAPPING WITH COURSE OUTCOMES (1- Weak Correlation; 2- Medium correlation; 3- Strong Correlation)					
	CO-1	CO-2	CO-3	CO-4	CO-5	CO-6
U6-O1	1	-	1	1	-	3
U6-O2	1	1	1	1	-	3
U6-O3	-	-	-	-	-	3
U6-O4	-	-	-	-	-	3
U6-O5	-	-	-	-	-	3

6.1.Concept of Green Buildings

The creation and maintenance of buildings have a significant direct and indirect influence on the environment. Large quantities of energy and natural resources are used by buildings, along with trash and environmentally hazardous pollutants. In previous decades, typical structures were major consumers of both fossil fuels, which release greenhouse gases, and chlorofluorocarbons (CFC), which weaken the ozone layer. To satisfy the expectations for new and rebuilt buildings that are safe and healthy while reducing negative environmental consequences, designers and builders confront a growing challenge as economies and populations grow. Every industry has the ability to contribute to the drive for more productive, energy-saving business and production processes, particularly in the building industry.

Buildings already account for more than 30 percent of India's electricity use, and two-thirds of the buildings that will exist in India by 2030 are yet to be built [1]. There is a huge waste of energy in such buildings due to a lack of awareness about the efficient use of energy, untrained manpower, or just negligence. There is a substantial consumption of electrical energy in buildings, especially considering Indian scenarios like air conditioners, lighting, computers, fans, motors for water pumping, etc. The majority of these structures utilize energy inefficiently, produce a lot of trash during their creation and usage, they release a lot of pollutants and greenhouse gases into the atmosphere. Unlike traditional structures, green buildings aim to utilize land and energy effectively, preserve water and other resources, enhance the quality of the air both inside and outside, and use more recycled and renewable materials. Green buildings still make up a very small portion of all existing structures, but their numbers are growing quickly.

According to Environmental Protection Agency (EPA) defines green building as follows [2], "***the practice of creating structures and using processes that are environmentally responsible and resource-efficient throughout a building's life-cycle from siting to design, construction, operation, maintenance, renovation and deconstruction. This practice expands and complements the classical building design concerns of economy, utility, durability, and comfort. Green building is also known as a sustainable or 'high performance' building.***"

A few keywords associated with green building and its material resource efficiency are

- ***Life Cycle Assessment (LCA)***

  It is a method used to perform complete environmental accounting of the impact of the building considering the material used in its construction.

- ***Life Cycle Cost Analysis (LCCA)***

  This measures the actual financial profits of a certain particular material, its life span, replacement cost, maintenance cost, and performance on the basis of economics.

- ***Embodied Energy Analysis***

  Embodied energy is the complete energy used for the creation and transportation of the materials used in buildings. This energy usually accounts for 15% to 20% of a building's total energy use during a 50-year period. Thus, having low embodied energy implies a reduced life cycle carbon footprint.

Some of the key features of the green building are given in Fig. 6.1.

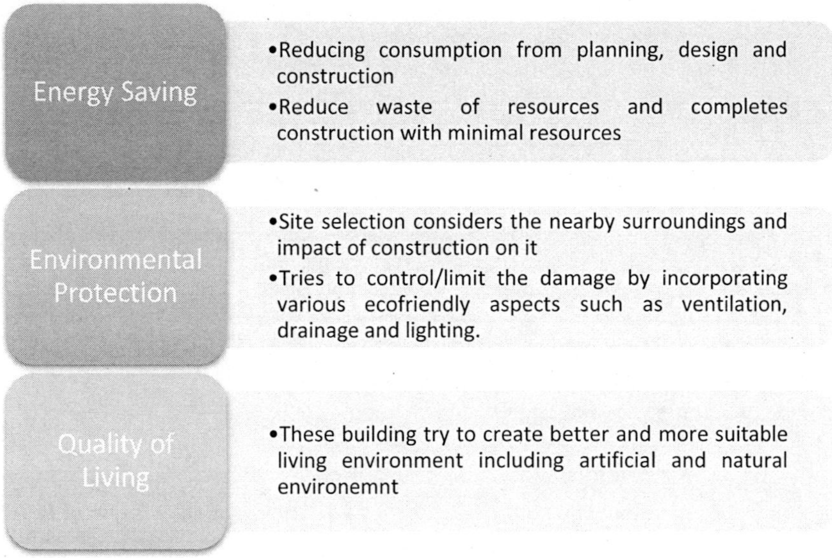

Fig. 6. 1. Salient features of green buildings

6.1.1. Material for Buildings

The green building materials are chosen on the basis of specific criteria. The various objectives of green buildings are the driving factor for the material selection, and these objectives are shown in Table 6.1. There is the different life cycle of phases associated with the building material throughout the life cycle of the buildings, which is depicted in Fig. 6.2.

Table 6.1. Key objectives of green buildings to select appropriate building material

Process	Objectives	Description
Manufacturing Process (Pre-Building Phase)	Waste Reduction	To make the production process more efficient by reducing the amount of scrap material.
	Pollution Prevention	Reduced the air, water, and soil pollution associated with the manufacturing process
	Reduced Embodied Energy	Revision to a manufacturing process that saves energy reduces the embodied energy of the material

	Recycled	Incorporation of waste materials from industrial processes or households into usable building products
	Natural Materials	Use natural materials such as wood as they are nontoxic and less processing to the environment
Building Operations (Building Phase)	Energy Efficiency	Reduce the amount of artificially generated power that must be supplied to the building site
	Water Treatment and Conservation	Reducing the amount of water that must be treated by municipal septic systems
	Nontoxic	Avoid the usage of adhesives and other materials that emit dangerous fumes and are toxic to workers and occupants
	Renewable Energy Source	Replace traditional building systems that promote daylighting and passive solar heating and are dependent on the off-site production of energy
	Longer Life	Use materials with longer life to prevent frequent replacement
Waste Management (Post-Building Phase)	Biodegradable	Potential to naturally decompose when discarded
	Recyclable	To be used as a resource in the creation of new products
	Reusable	Can be easily extracted and utilized in other new buildings once the building is decommissioned

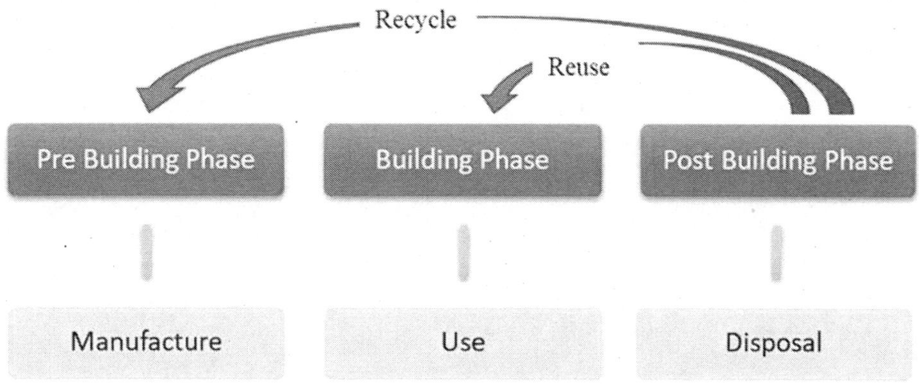

Fig. 6. 2. Three phases of the building material life cycle

Some of the green building materials commonly used are

- Fly Ash Concrete
- Natural clay
- Corrugated metal
- Wheatboards
- Plywood
- Recycled polyethylene terephthalate (PET)
- Natural cork
- Wool and wool blends
- Fiberglass batts
- Limestone
- Steel
- Aluminum
- Bricks and Tile

6.1.2. Designing and Building Orientations

A green building is a long-lasting, practical, and creative approach toward sustainability. These buildings incorporate knowledge, tradition, and a collaborative design process with modern science and technology. Building a solid green building project team with design expertise is essential for design success. Designers with experience in green construction methods, such as solar design and sustainable site planning, include experienced house designers, architects, landscape architects, and interior designers. They are able to develop a vision that takes the project's objectives and financial constraints into account. The designing of a green building involves the following:

- Proper selection of site location
- Environment suitability
- Cost-effective infrastructure
- Affordable and compact design
- Proximity to basic services
- Walkable neighborhoods
- Incorporate passive solar design
- Reduce the heat island effect
- Proper building orientation
- Passive cooling (using false ceilings)

One of the main aims of the design of green buildings should be making the building more affordable, and to achieve that, the following steps are taken:

- Designing buildings with smaller footprints
- Optimizing the size and orientation of the building
- Natural daylighting provisions
- Square dimensions are more resource efficient than rectangular ones
- Eliminating hallways from the floor plans

Environmental consciousness is the foundation of the contemporary global construction paradigm and also the green building concept, which necessitates the use of sustainable building orientation and construction methods. Sustainability can easily accommodate in the built environment by a basic component like orientation depending on terrain, angle of exposure, and climate. Enhancing the building's sustainability may be done in large part by controlling how the sun and wind interact with its façade. A building receives extra points from the Indian Green Building Council (IGBC) for not having to rely extensively on thermostats and air conditioners to manually control the temperature. Let's look at the many construction components you need to take into account, in accordance with IGBC guidelines, to effectively utilize solar gains (for more information about IGBC, please scan):

- ***East/West axis alignment***

The primary façade of the building should be oriented within a 15-degree range of either the eastern or western axis. Green buildings are intended to absorb as much sunlight as possible; hence, they would ideally be rectangular, with the east and west faces shorter than the northern and southern ones. The southern half of the earth receives the sun for the majority of the day.

- ***Southern exposure***

  To allow south-facing windows to capture solar heat during even the coldest winters, the building's southern face should be positioned within 20 degrees on each side of the true south. The heating and cooling of the interiors of the buildings would deviate further with each degree over the 20 degrees of true south.

- ***Surface areas of windows***

  The biggest windows in a structure must be those facing south; ideally, their surface areas should be around 50% more than those facing east and west combined. Overhangs and other architectural elements should assist the south-facing windows in dissipating some of the summertime noon heat. The lowest surface area windows should be those facing west since they are overexposed to intense solar heat during summer sunsets and are of little use for capturing solar heat in the winter.

- ***Orientation with respect to topography***

  Due to the gradient and geography, the orientation of the building's location is crucial in determining the angle of incidence of radiation from the sun. The incident solar radiation is minimized in the summer and increased in the winter if the landscape slopes. As a result, the site selection should take into account the latitude, the amount of incoming solar radiation, and the slope of the ground. A southern slope is the hottest during the winter since it directly faces the winter sun.

6.2.LEED Ratings

LEED stands for Leadership in Energy and Environmental Design. LEED is a credit-based certification system developed by the United States Green Building Council (USGBC) to certify green and sustainable buildings. LEED-based green design has the following positives:

- Improved public and environmental health
- Reduced operating costs
- Promoted building and organizational marketability
- Improves occupants' productivity

➢ Create a sustainable future

Fig. 6.3 depicts the performance of the LEED rating system in six key areas of human and environmental health.

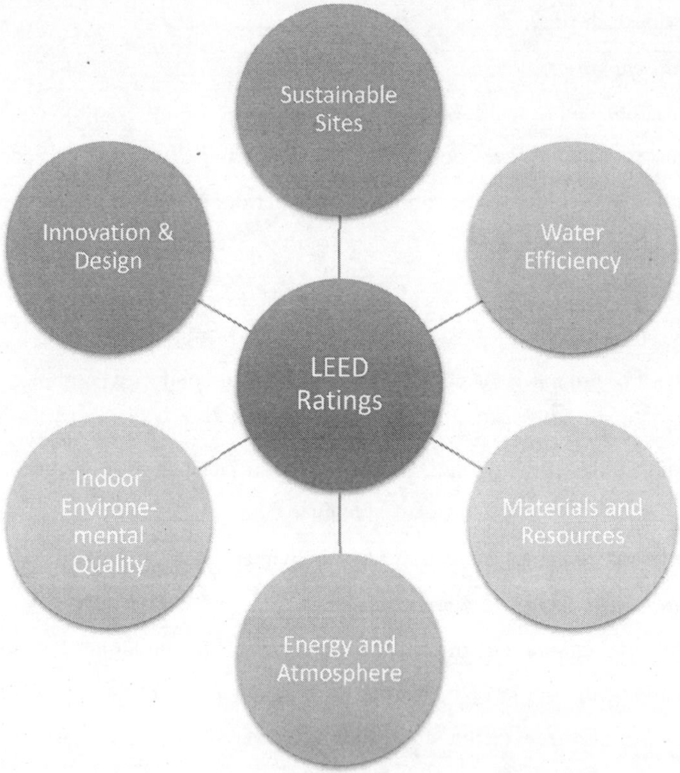

Fig. 6. 3. Key performance areas recognized by LEED certification

LEED certification shows the commitment to sustainability by meeting the highest performance standards. LEED-certified buildings and sites have the following benefits:

➢ Utilization of important resources efficiently
➢ Contribute to a healthier work environment
➢ Promote productivity and comfort
➢ Improves assets value over time
➢ Promote new building innovation, materials, and equipment
➢ Qualify for government and international incentives
➢ Establish leadership in building industry and marketplace

LEED certification is given after the successful submission of the application compliance with the necessary requirements of the rating system, along with the registration and certification fees. Points received, pending, and rejected are noted in a Preliminary LEED Review document released by the USGBC. After receiving the preliminary evaluation, the project team has 30 days to make adjustments and/or additions to the application. Within 30 days of receiving the resubmission, the USGBC performs a Final LEED Review of the application and proposes a final application score to the LEED Steering Committee. Within two weeks of receiving the suggested final application score, the LEED Steering Committee approves or rejects it and notifies the project contact of the LEED Certification. The project team has 30 days after receiving notice of the LEED certification to accept or dispute the designation. The LEED Certification is final after the project has been approved or if an appeal against the rating has not been filed within 30 days. A LEED Certified Green Building may then be used to describe the undertaking. A certificate and a metal LEED plaque denoting the certification level are given to the project team by the USGBC. Each appealed credit is subject to a $500 charge. The certification is provided on the scoring of the above-mentioned categories in Fig. 6.3. These scores/credits determine whether the buildings qualify for certification. A total of 69 credits are given, depending on the score obtained following certification is provided:

- Certified: 26 to 32 credits
- Silver: 33 to 38 credits
- Gold: 39 to 51 credits
- Platinum: 52 to 69 credits

Further, the breakup of LEED rating system credits are mentioned in Table 6.2.

Table 6.2. LEED Rating System

LEED Categories	Points	Contribution
Sustainable Sites	14	20%
Water Efficiency	5	7%
Energy and Atmosphere	17	25%
Materials and Resources	13	19%
Indoor Environmental Quality	15	22%
Innovation and Design Process	5	7%
Total	69	100%

There are 10 different LEED Rating programs for different types of constructions, namely

- New Construction
- Existing Building
- Commercial Building
- Core and Shell
- Schools
- Retail-New Construction
- Retail and Commercial Interiors
- Healthcare
- Homes
- Neighborhood Development

6.3.Identification of Energy Related Enterprise

Modern society consumes a large amount of energy globally. The demand for energy is growing day by day. We cannot imagine, the growth of a country and the life of a person without the use of energy resources. Therefore, energy related sectors are one of the most important sectors of any country. The energy industry includes companies involved in the exploration and development of oil or gas reserves, oil or gas drilling, coal mining, nuclear fuels, generating power, manufacturing different components used in power plants, etc. The energy sector includes companies that are primarily involved in the business of supplying energy, such as fossil fuels or renewable.

The companies within energy sectors are classified as companies based on how the energy that they produce is sourced. They may be classified as non-renewable and renewable energy sources.

Non-renewable

Petroleum and oil

Natural gas

Diesel

Coal

Nuclear

Renewable

Hydro power

Solar

Wind

Biofuels

The energy industry also includes secondary sources such as electricity. Generation, transmission, and distribution of electricity and all related companies involved in the manufacturing of different parts used in energy-related activities. In 2021, in the USA, petroleum was the most consumed energy source at 36%, followed by natural gas at 32%, renewable energy sources at 12%, coal at 11%, and nuclear at 8%. In India, coal is the country's top energy source, with a share of 44% in 2021, followed by oil (24%) and biomass (22%). Natural gas covers 6%, and primary electricity (hydro, nuclear, solar, and wind) is 4%.

Role of energy sector:

The energy sector plays a crucial role in the economy of a country. Apart from powering homes, transport, and factories, energy sources are also a component of various companies. Energy sectors may also be classified as energy equipment & services, Oil, gas, and consumable fuels.

Various subsectors are:

Oil and gas drilling

Oil & gas equipment & services

Oil & gas exploration and production

Oil and gas storage & transportation

Coal and consumable fuels

Mining of nuclear fuels

Utility companies, on the other hand, provide electricity, water, and other public utilities to their consumers. There are also many energy related exchange traded funds that can give retail investors exposure to the energy industry. The energy industry is more diversified and extensive than merely the oil and gas industry. In the future, it is expected that renewable energy sources have the potential to play an important role as the demand for electric vehicles continues to grow.

The energy industry comprises:

- The fossil fuel industries, which include petroleum, coal and the natural gas industries.
- The electric power generation, transmission, distribution and sales industries.
- The nuclear power generation associated companies.
- The renewable energy sectors
- Corporates involved in selling of energy and its products.

Top 10 Energy Companies in India

1. Adani Power Ltd.
2. CESC Ltd.
3. Gujarat Industries Power Company Ltd.
4. Indraprastha Power Generation Company Ltd.
5. NHPC Ltd.
6. NTPC Ltd.
7. Nuclear Power Corporation of India Ltd.
8. Power Grid Corporation of India Ltd.
9. Rattan India Power Ltd.
10. Tata Power Company Ltd.

World top Energy companies by Market Capitalization

Saudi Aramco	Saudi Arabia
ExxonMobil	United States
Chevron Corporation	United States
Shell	United Kingdom
Total Energies	France
NextEra Energy	United States
ConocoPhillips	United States
PetroChina	China
Equinor ASA	Norway
BP	United Kingdom

6.4. Embodied Energy Analysis

The majority of the time, embodied energy analysis is more concerned with the energy used to sustain a consumer, and therefore, all energy depreciation is attributed to the user's end demand. To determine the amount of energy that is incorporated in the goods and services provided by nature and human civilization, many approaches employ various data sizes. International agreement on the suitability of data scales and methodology is still awaited. For any particular material, the embodied energy levels might vary greatly depending on this challenge. Embodied energy calculations may omit crucial information if there is no comprehensive global embodied energy public dynamic database. Examples of this include the rural road/highway construction and maintenance required to move a product, marketing, advertising, catering services, non-human services, and the like.

The embodied energy is the energy needed to build and operate the structure. For instance, the energy needed to produce the bricks, transport them to the construction site, lay them out, plaster them, then (if necessary) paint and replaster them over the wall's lifespan. Best practices would also incorporate recycling and demolition energy estimates. Embodied energy in residential structures generally makes up between 30 and 100 percent of the overall life cycle energy consumption [3]. However, the significance of embedded energy is typically more than these numbers suggest. The units of energy used to describe embodied energy are typically MJ or GJ. The amount of primary energy used to power the resource processes, as well as its effectiveness, will determine how this corresponds to carbon emissions. For instance, an aluminum product from a smelter powered by hydro power will have a high energy content, but relatively low embodied carbon. The intention is to cite both values, although it should be stressed that an exact estimate of carbon emissions is currently more challenging to make and more variable. Only the supplied material data can guarantee the accuracy of the embodied energy model.

Most embodied energy figures for specific materials are quoted using a "cradle to gate" boundary. Consumers must also consider transport, assembly, maintenance, and demolition components of embodied energy. In addition, care should be taken to ensure that primary energy consumption is calculated, not delivered energy (which will understate the real energy cost). A stepwise breakdown of embodied energy analysis is shown in Fig. 6.4. The cradle to gate method, or all the energy necessary to get the product to the factory gate so it can be transported to the building site, often defines the most significant components of embodied energy. The complexity of embodied energy might be quite high, even within a "cradle to gate" computation.

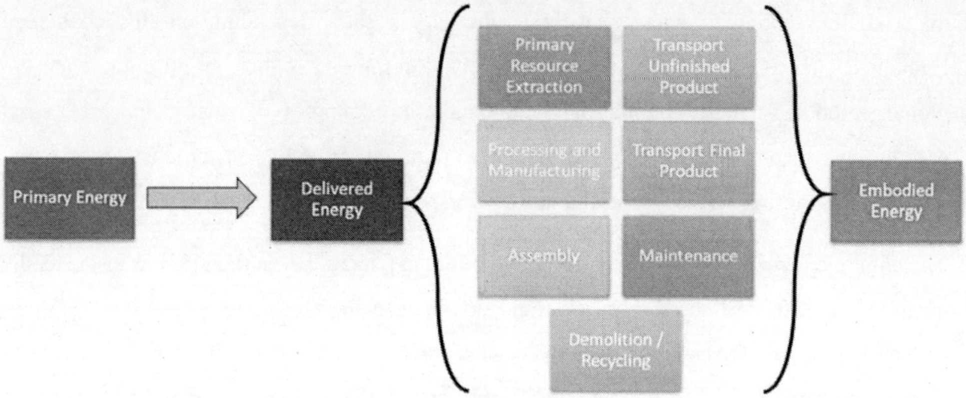

Fig. 6. 4. Breakdown of embodied energy calculations

The embodied energy analysis and a building's environmental effect are mostly done using the SBTool, UK Code for Sustainable Homes, and USA LEED. Because there are so many factors to consider, scientists have not yet established exact universal numbers for the idea of embodied energy, but most believe that items may be contrasted to determine which has more and which has less embodied energy. Energy Returned on Energy Invested (EROEI) provides a baseline for calculating embodied energy.

$$Embodied\ Energy = Final\ Energy \times \frac{1}{EROEI - 1} \quad (1)$$

Given an EROEI amounting to eight, e.g., a seventh of the final energy corresponds to the embodied energy.

Some Examples regarding embodied energy are given in Table 6.3.

Table 6.3. The embodied energy content of a few items

Item Description	Embodied Energy
Volkswagen Golf A3 with petrol engine	18000 kWh = 65.4 GJ
Volkswagen Golf A4 with Turbo charged petrol engine	22000kWh = 81.75 GJ
Road construction	1/18 of fuel consumption
Buildings	Approx 30% of all energy consumed in lifetime
Fuels	14.3% of the fuel EROEI consumed

For more information and to download the data, please scan

6.5. Tools for Measuring Sustainability

Industry advancement is now primarily measured by sustainability, and effective company strategies are taking this into account. Advancement in operations, assessing performances, monitoring progress, and process evaluation are just a few of the demands for sustainability assessment. There are certain steps and framework to be followed to develop a proper sustainability factor. This framework is presented in Fig. 6.5.

1. Defining a system
   - A proper system boundary is drawn for further analysis.
2. Elements of the system
   - The working conditions, process parameters and characteristics are defined in this step.
3. Selection of indicators
   - The indicators is selected of which measurement has to be done.
4. Measurement and assesment
   - Proper assessing tools are used and tests or experiments are performed for the pre-defined indicators to give a value for the indicators measurement.
5. Analysis and review
   - Proper analysis and interpretation is done and tools are used to improve and revise the processes present in the system

Fig. 6. 5. Framework for sustainability measurement

Sustainability is measured on the basis of four categories; these are environmental care, natural depletion, societal wellbeing, and economic wellbeing [4]. The summary of the assessment tools for the measurement of sustainability is presented in Table 6.4.

Table 6.4. Various sustainability assessment tools

Assessment Tool	Summary
Cascadia Scorecard [5]	An easy-to-use evaluation instrument that has to be reviewed annually to identify changes in seven themes that are relevant to a particular geographic area. The evaluation tool might be used to determine a housing project's long-term viability even though it was designed to demonstrate patterns in the US Pacific North West. This tool's usage of data that is widely accessible is one of its benefits. It was developed in the USA.
One Planet Living (OPL) *For more information, please scan*	OPL offers a blueprint that may be used globally to handle the diversity of sustainability issues. The use of performance metrics, goal-oriented indicators, and aggregate indices is combined. There are ten different topic sections. Targets are given for short-, medium-, and long-term time spans. OPL makes it possible to use presently accessible measurements to address some broad sustainability categories, while in other cases, it necessitates the development of location-specific data. This tool was developed in the UK.
Sustainable Project Appraisal Routine (SPeAR) *For more information, please scan*	This framework was developed by the UK. Reviewing and maximizing sustainable opportunities is the aim. SPeAR has the potential to address several sustainability-related issues. 22 thematic categories often employ performance measurements; however, these can be increased or decreased based on significance. Since the units of measurement are so broad, they might skew how sustainable success is seen (for instance, "great" is defined as "targeted or inventive" or "holistic" or "gaining an award").
VicUrban Masterplanned	This evaluation tool offers a fundamental foundation to aid in the design and creation of sustainable communities in new mixed-use

Community Assessment Tool	complexes with at least 500 homes. It typically employs 11 theme-based performance metrics. This tool was developed and put forth by Australia.

The above tools utilize indicators to inform public and policy makers of the sustainability governance process. These indicators provide in-depth knowledge of and the relationship between environment and socio-economic growth. Some of these indicators are listed below:

- ***Environmental Sustainability Indicators***
  - Global Warming Potential
  - Ozone Depletion Potential
  - Waste Treatment
  - Aerosol Optical Depth
  - Energy Resources
- ***Economic Indicators***
  - Gross Domestic Product
  - Trade Balance
  - Local Government Income
  - Profit, Value and Tax
  - Investment
- ***Social Indicators***
  - Equity
  - Education
  - Housing
  - Community Cohesion
  - Social Security
  - Employment Generated
  - Health and Safety

Summary

In this unit, the concept of energy conservation and its importance in buildings is discussed. The concept of green building and green architecture, the structural design aspects, and the materials required to make an energy-efficient building are discussed. Key performance areas recognized by LEED certification, LEED rating programs for different types of constructions, and benefits to LEED-certified buildings and sites have been described. The embodied energy and its analysis, which is the energy needed to build and operate the structure, is also discussed in this unit. The sustainability which is measured on the basis of four categories, environmental care, natural depletion, societal wellbeing and economic wellbeing are discussed and the summary of the assessment tools for the measurement of sustainability is presented.

Short and Long Answer Type Questions

1. List a few key words associated with green building and its material resource efficiency.
2. What are the objectives of green buildings? On what specific criteria, the green building materials chosen?
3. List some of the green building materials commonly used.
4. What are the important criteria involved in the designing of a green building?
5. What are the key performance areas recognized by LEED certification?
6. Describe the concept of daylighting.
7. Write a short note on embodied energy analysis.
8. On what basis the sustainability of building from energy point of view is measured?
9. What makes products green? How can I verify the manufacturers' performance claims?
10. What questions are to be asked when taking on a green building project?
11. What can be done to make sustainability and green building cost-effective?
12. What does the future of sustainable building look like?
13. How is energy-efficient building affecting interior air quality?
14. How do you introduce and apply green methods in remote areas with weak economies and basic infrastructure?
15. What is the most effective way to enhance people awareness toward energy consumption in their buildings?

References and Suggested Readings

1. S. Itkelwar, D. Tewari, B. Natarajan, and S. Kumar, "Residential Energy Data Management: Alliance for an Energy Efficient Economy," Shakti Foundation, Delhi, 2020.
2. Green Buildings, Environmental Protection Act
3. R.H. Crawford, R.J. Fuller, G.J. Treloar, B.D. Ilozor, "Embodied energy analysis of the refurbishment of a small detached building". In Proceedings of 36th Annual Conference of the Australian and New Zealand Architectural Science Association (ANZAScA): The Modern Practice of Architectural Science from Pedagogy to Andragogy, 2002.

4. E. Karol, and J. Brunner, "Tools for Measuring Progress Towards Sustainable Neighborhood Environments," Sustainability, vol. 1, pp. 612-627, 2009.
5. Northwest Environment Watch, "Cascadia Scorecard: Seven Key Trends Shaping the Northwest," Slightline Institute, Seattle, US, 2004.

****

7 Engineering For Energy Conservation

(Energy Auditing, Efficiency & Energy Conservation)

UNIT SPECIFICS

In this unit, the following topics have been discussed for basic understating related to energy auditing, energy efficiency, and energy conservation:

- Concept of energy auditing and need for energy auditing
- Different types of energy auditing
- Case studies on energy auditing
- Concept of energy efficiency

Here, in this unit the main focus is to gain knowledge about the conservation of energy. In the present scenario, the demand of energy is increasing on day-to-day basis. To meet energy demand, a lot of measures at individual, society, and government levels are required. Generated energy should not just meet our needs, but simultaneously it should also not have any adverse impact on our environment. Energy has also become as essential as air and water. Therefore all efforts should be made to conserve energy. Energy auditing is one of the methods that suggest how we can use energy efficiently. Energy auditing methods and a few case studies are presented in this unit to understand the basic concept of efficient use of energy. An energy audit helps us to understand how the energy is used and helps in identifying areas where the waste of energy can occur and where the scope for improvement exists.

RATIONALE

This unit introduces the concepts of energy auditing and energy efficiency in the context of energy conservation. Energy conservation and efficiency may be related, but they have distinct definitions in the energy world. Energy conservation means a reduction in energy consumption but without making any sacrifice in the quality or quantity of production. Energy conservation involves using less energy by adjusting your behaviors and habits. Energy efficiency, on the other hand, involves using technology that requires less energy to perform the same function.

PRE-REQUISITES

Basic Knowledge of Environmental Science and Energy

UNIT OUTCOMES

List of outcomes of this unit is as follows

U7-O1: To understand the basic concept of energy efficiency and energy conservation.

U7-O2: To understand the basics of energy auditing.

U7-O3: To classify and identify the types of energy auditing.

U7-O4: To know the best practices for efficient use of electrical energy.

U7-O5: To analyze the different factors responsible for energy management and its economics.

Unit outcomes	EXPECTED MAPPING WITH COURSE OUTCOMES (1- Weak Correlation; 2- Medium correlation; 3- Strong Correlation)					
	CO-1	**CO-2**	**CO-3**	**CO-4**	**CO-5**	**CO-6**
U7-O1	1	-	2	-	-	3
U7-O2	-	-	1	-	-	3
U7-O3	-	-	-	-	-	3
U7-O4	1	-	2	-	-	3
U7-O5	1	-	2	-	-	3

7.1. Energy Auditing

Energy auditing is a judicious and effective use of energy to maximize profits and improve competitive positions. This involves inspections, surveys, and analysis of energy flows for the identification of energy saving opportunities in a built infrastructure. In other words, it can be said that it is a process implemented in a building or a system to reduce the amount of energy input into the system without adversely impacting the output. Energy Auditing is a step-by-step approach for decision making in the field of energy management. As per the Bureau of Energy Efficiency India, the comprehensive definition of energy auditing is similar to energy management as follows [1],

"The strategy of adjusting and optimizing energy, using systems and procedures so as to reduce energy requirements per unit of output while holding constant or reducing total costs of producing the output from these systems"

As per the Energy Conservation Act, 2001, Energy Audit is defined as "the verification, monitoring and analysis of use of energy including submission of technical report containing recommendations for improving energy efficiency with cost benefit analysis and an action plan to reduce energy consumption".

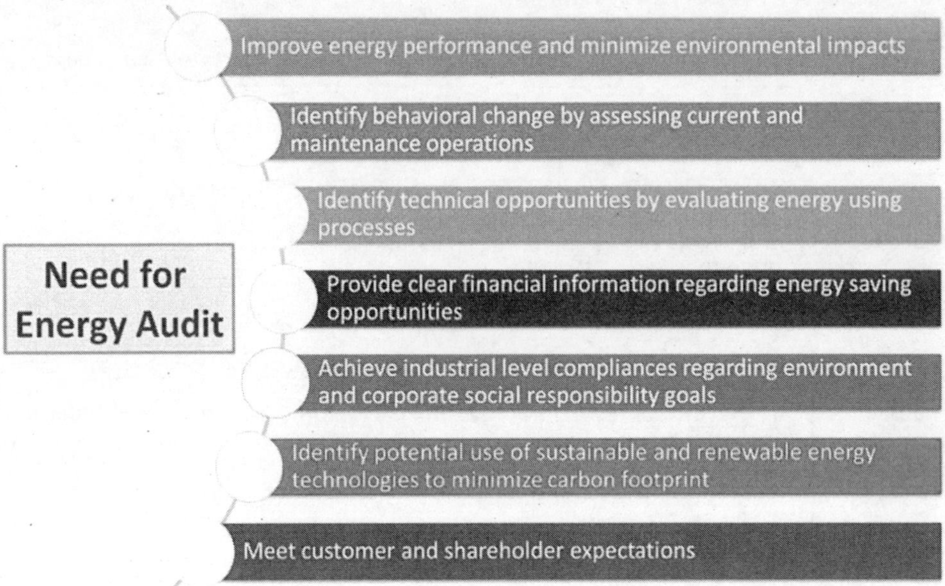

Fig. 7. 1. Need for Energy Auditing

Energy analysis and auditing would provide a helpful direction for operations that are essential for manufacturing and utility operations, such as reducing energy costs, preventative maintenance, and quality control. It will assist in maintaining attention to fluctuations in energy prices, the availability and dependability of the energy supply, choosing the right energy mix, identifying energy-saving technology, retrofitting for energy-saving equipment, etc. The requirement of energy auditing is depicted in Fig. 7.1.

7.2. Types of Energy Auditing

There are different types of energy audits that can be performed. These depend on the following factors such as

i. Type of industry
ii. Functioning of industry
iii. Depth of the final audit that is required
iv. Potential benefits desired from the audit

The auditor will identify the energy-consuming areas and create a comprehensive list of energy-saving strategies. Thus, after careful assessment of the above factors, three types of energy audits can be performed [2], namely

i. Preliminary Energy Audit
ii. Targeted Energy Audit
iii. Detailed Energy Audit

7.2.1. Preliminary Energy Audit

A preliminary energy audit is also known as a 'walk through energy audit', where a simple survey is conducted by the auditor to investigate energy consumption of the organization. It is the easiest and shortest form of audit and the initial stage in the auditing process. It is concentrated on the main area of energy usage. This type of energy audit requires significantly lesser time span. The qualified energy manager, certified auditor, or professional engineer completes the walkthrough audit. Finding an energy-saving solution requires more information than this type of audit can provide. The auditor can operate with high precision and profitability in less time. This kind of energy audit provides the auditor with all the information about an organization that is required for a more thorough energy audit. The scope of the preliminary energy audit are:

- Evaluate the energy consumption and scope for saving
- Minimal interaction with the organization personnel
- Identify the most likely and easiest areas of concern
- Identify immediate (especially/low cost) improvements
- Analyze pre-existing data or easily obtained data
- Develop a brief layout for an in-depth/detailed study

7.2.2. Targeted Energy Audit

The findings from the preliminary energy audit report are utilized in a targeted energy audit. This type of energy audit is focused on a specific and certain objective of the project. The approach and methodology are variable and depend on the type of structure and industry the audit is done. The outcomes are the recommendations provided by the auditor that need to be implemented for energy saving. This report is concise, clear, and includes a plan of action for improving energy efficiency. The scope of this type of audit includes:

- Minimal interviews with organization personnel

- Relate energy consumption with the type of organization
- Specific area is to be covered during this audit
- Find methods for improving energy efficiency and conservation
- Suggest the easy ways to achieve the goals

7.2.3. Detailed Energy Audit

This audit is often referred to as an extensive or investment-grade energy audit. A thorough investigation of potential capital-intensive improvements, including modeling and simulation, may be required for the audit of the detailed energy. For high accuracy, a complete energy audit may require data gathering over a lengthy period of time. It will produce precise energy consumption records and detailed savings reports, making it simple for contractors to understand what specific solutions to implement. Since it assesses all significant energy-using systems, a thorough audit offers a business a clear strategy for implementing an energy project. The most precise calculation of energy savings and costs is provided by this kind of audit. It measures the energy consumption of all significant equipment, takes into consideration the interacting impacts of every project, and provides thorough projections for energy cost savings and project costs. The energy balance is one of the crucial components of a thorough audit. This is based on measurements of energy usage, an inventory of systems that utilize energy, and assumptions about how things operate right now. The costs on a utility account are then compared to this projected utilization. The detailed energy audit is carried out in three phases: pre-audit phase, audit phase, and post-audit phase.

- **Pre-Audit Phase**

The energy auditor organizes and arranges the energy audit during the pre-audit stage. A formal interview is held with the Chief Executive Officer (CEO), plant manager, energy manager, and production manager at this phase. In this step, a brief meeting is also arranged with all kinds of heads and the affected person. The energy auditor or qualified energy auditor plans to do the following actions during the pre-audit phase. A pointwise summary of the actions performed by energy auditor in pre audit phase are:

- Discussion with senior management
- Discussion on guidelines related to the economy of the organization
- Analyze consumption patterns with concerned personnel
- Collect site drawings and layout of major components
- Complete tour of the premises with the site engineer

- **Audit Phase**

This phase is the next stage of a comprehensive energy audit. This phase entirely depends on the simplicity and the nature of the site, as it can take from days to months. This audit phase considers working hours as well as non-working hours and nights to ensure consideration of every detail and aspect in every sense. This stage of detailed energy audit performs surveys and collects data of each energy consuming and generating system. Thus, gathering in depth knowledge about the organization or specific site or a department. To make sure nothing is missed, checks of plant operations are made whenever feasible over long periods of time, including evenings, weekends, and regular working hours. During this phase an auditor carries out the following actions:

- Collect data about the different sources of energy powering the site
- Analyze the previous electricity bills and tariff data
- Collect information about the loads
    - The information about load is critical in determining accurate energy auditing report. The load information generally includes the type of loads, ratings of the loads, power factor and current intake of the loads, the time of usage of loads, purpose of loads, etc.
- Review and propose changes in the existing energy management program
- Energy flow diagram

- **Post Audit Phase**

For the evaluation of a capital budgeting decision, a set of procedures must require after the fact. The plan of action for the post-audit phase is implementation and follow-up. The result is to assist and implement energy conservation (ENCON) recommendation measures and monitor the performance.

7.3. Instruments required for energy auditing

To measure the energy consumption, quantification and identification for an energy audit required different instruments. These instruments should be portable, durable, easy operability and relatively inexpensive. These instruments mostly involve the measurement of electrical quantities such as voltage, current, power factor, active power, reactive power, demand, energy consumption, frequency, etc. In addition to these other important parameters are temperature, heat flow, radiation, air and gas flow, revolutions per minute, total dissolved solids, pH, relative humidity, flue gas analysis, etc.

Table 7.1. Key instruments required for specific types of systems for energy audit [3].

Electrical Systems	Voltmeter
	Power meter
	Ammeter
	Multimeter
Temperature Measurement	Surface Pyranometer
	Portable Electric
	Thermometer
	Thermocouple Probe
	Infrared Thermometer
	Infrared Camera
Steam System/Compressed Air	Ultrasonic Leaks
	Detectors
	Steam Trap Tester
HVAC Systems	Manometer
	Psychrometric
	Anemometer
Building	Light Meter
	Measuring Tape
	Thermal Image
	Camera
	Lux meters
Data Loggers	4-20Ma Loggers
	0-10V Loggers
	Digital Loggers
	Vibration Loggers
	Light Loggers

7.4. Software used for Energy Auditing:

Simprosys: It is helpful for mass transfer, design projects, courses in unit operation, heat, and mass balance calculation.

Road pollution: It is helpful for the installation of road lighting and the assessment of the environmental impact due to light pollution.

MATLAB: MATLAB is an advanced simulation software where any job such as automation, lighting, layouts, engineering, and mathematics functions can be performed.

Piping System Fluid Flow: Piping system fluid flow can be used for accurate simulation of network performance before a network design or modified a network

IHAT: Indoor humidity assessment tool (IHAT) software is used to check the humidity level in a building, which includes the use of an energy recovery ventilation (ERV) system.

eQUEST: Energy Quick energy simulation (eQUEST) software is a user-friendly and freeware tool that is used to predict energy use, building layout, HVAC requirement, hourly weather data, and energy cost.

DIALux: DIALux is used for advanced lighting study and energy efficiency measures for roads, areas, buildings, and offices.

Easy Audit: Easy Audit doesn't require internet access. It is designed for use on tablets with a stylus but runs on any computer or device that runs MS Excel.

PSAT: The pumping system assessment tool (PSAT) is used for the measurement of pump efficiency and pressure drops.

SSAT: Steam system assessment tool (SSAT) is used to measure the performance of the steam system

7.5.Economic Analysis

Economic viability frequently ends up being the deciding factor for management approval. Numerous techniques might be used to carry out the economic analysis. As an illustration, consider the Pay Back Method, Internal Rate of Return Method, and Net Present Value Method. Payback is often enough for minimal investment, short-term actions with appealing economic viability that use the simplest of techniques [4]. Fig. 7.2. shows the worksheet for economic feasibility.

Investment	Annual Operating Cost	Annual Savings
• Equipements • Cicil works • Instrumentation • Auxiliaries	• Cost of Capital • Maintenance • Manpower • Energy • Depreciation	• Thermal Energy • Electrical Energy • Waste Disposal • Raw materials

Fig. 7. 2. Parameters for the economic feasibility of the energy audit

Fixed cost

- Cost which is dependent on maximum demand and units generated

Semi fixed cost

- Charges levied on the installed capacity of the plant and are independent of the elctrical energy output of the plant

Running cost

- These cost depend on the number of hours the plant is in ooperation or upon the number of units of electrical energy generated.

Fig. 7. 3. Different types of costs of electrical energy

The cost of procurement and electricity prices for different states in India vary due to geographical conditions, consumer base, etc. Hence, the following factors are considered for energy management and its economics

- Price at source and transport cost
- Transport type
- Quality of fuel
- Calorific value of fuels
- Energy charges
- Time of use charge

- ➢ Peak demand charges
- ➢ Power factor charges
- ➢ Tariff rates
- ➢ Slab rate cost and its fluctuations

Table 7.2. represent a brief description about the various types of tariffs incurred for electrical energy. Tariffs include the cost of investment, operation, and equipment on a particular project.

Table 7.2. Types of tariffs (For further information, scan)

Type	Description
Flat demand tariff	It is the charge paid by the consumers. Energy charges = Rs A. x where x is load connected in kW and A is the flat rate per kW
Simple tariff	This is used when there is a fixed rate per unit of energy consumed. $\text{Cost per kW} = \frac{\text{Annual fixed cost} + \text{Annual operating cost}}{\text{Total units supplied per year}}$
Flat rate tariff	Types of consumers are grouped and are charged at different rates. This considers load factor and diversity factor.
Step rate tariff	The cost of 1kWh is not fixed but decreases in steps as the energy consumption increases.
Block rate tariff	When given amount of energy is charged at a specified rate and the succeeding lots of energy are charged progressively at reduced rate.
Two-part tariff	Charged on the basis of maximum demand of the consumer and the units consumed.
Maximum demand tariff	Applied to bulk supplies and large industrial consumers
Power factor tariff	Makes distinction between charge per unit to be covered from two types of consumers i.e. good power factor and bad power factor
Off-peak tariff	Charged during off peak hours and night.
Time of Day tariff	Specific times have been defined by the electricity regulatory bodies as peak hours and off-peak hours. The consumers will be charged more for energy consumption during peak hours and less during off-peak hours.

$$\frac{\text{Net Savings}}{\text{Year}} \text{ (Rs/Year)} = \text{Annual Savings} - \text{annual operating costs} \quad 7.1$$

$$\text{Payback period in months} = (\text{Investment/Net Savings/Year}) \times 12 \quad 7.2$$

$$\text{Production factor} = \frac{\text{Current year's produciton}}{\text{Reference year's production}} \quad 7.3$$

$$\text{Plant energy performance} = \frac{\text{Reference year} - \text{Current year}}{\text{Reference year's production}} \quad 7.4$$

Above are some of the equations used for calculating the economic feasibility of the project after the energy audit. Based on the type of energy audit and analysis of the site, the outcome may be defined in three main categories:

- Low cost - high return
- Medium cost - medium return
- High cost - high return

Normally the low cost - high return projects receive priority. Other projects have to be analyzed, engineered and budgeted for implementation in a phased manner. Table 7.3 represents the priority guideline of the project depending upon the techno-economic feasibility.

Table 7.3. Project priority guidelines

Priority	Economic Feasibility	Technical Feasibility	Risk
A: Satisfactory	Low cost – high return	Existing technology is adequate	No Risk, most feasible
B: May be	Medium cost – medium return	Updation in technology required	Minor Risk, may be feasible
C: Held	Medium cost – medium return	Existing technology is not adequate	Doubtful
D: No	High cost – low return	Major overhaul required	Not adequate

7.6. Case Study

To perform a detailed energy audit Table 7.4 shows a step-by-step methodology for conducting the energy audit.

Table 7.4. Methodology for detailed energy Audit

Phase 1 Pre Audit Phase

- **Step 1**
  - Plan & organise
  - Walk through audit
  - Informal interview with managers
- **Step 2**
  - Conduct brief meeting with heads

Phase 2 Audit Phase

- **Step 3**
  - Data gathering, process flow diagrams and energy flow diagrams
- **Step 4**
  - Perform survey and monitoring
- **Step 5**
  - Conduct detailed trials and experiments to assess load variations and boiler/furnace efficiency
- **Step 6**
  - Analyze energy usage
- **Step 7**
  - Identification and developement of ENCON opportunities
- **Step 8**
  - Cost benefit analysis
- **Step 9**
  - Reporting and presentation with top management

Phase 3 Post Audit Phase

- **Step 10**
  - Implementation and Follow up

Case study on an energy audit of household appliances is considered in this example. As in homes and buildings, the main power and energy consuming components are

- Lighting: Bulbs, Compact Fluorescent Lamps (CFLS), Light Emitting Diodes (LEDs) installed for indoor and outdoor illumination
- Cooling: Fans, Air Conditioners (ACs), Refrigeration units for cooling purposes
- Heating: Geysers, microwaves, induction tops for cooking and heating purposes

  Entertainment: Televisions (TVs), music systems, computers for everyday need
- Other Appliances: Mixers, washing machines, Juicers, etc

Calculation of energy consumption per day for a 2BHK flat with typical area of 900 ft^2

Device	Power Rating (W)	Quantity	Usage hours (hours)	Energy consumption (kWh)
Lighting				
Tubelight (LED)	18	05	10	0.9
Bulbs (LED)	8	10	05	0.4
Cooling				
AC	1750	01	06	10.5
Fans	70	03	06	1.26
Fridge	100	01	24	2.4
Heating				
Geyser	2000	01	0.5	1
Induction top	1200	01	0.5	0.6
Entertainment				
TV	25	01	02	0.05
Computer	200	01	02	0.4
Laptop	80	01	02	0.16
Miscellaneous				
Mixer	750	01	0.25	0.19
Washing Machine	400	01	1.5	0.6
Total Energy				18.46

Thus, the per month energy usage of the above considered scenario is 18.46 x 30 = 553.80 kWh

Considering the tariff rates as given below (these rates are inclusive of fixed per kW meter charges)

Units	Tariff Rates (Rs./kWh)
0-100	1.65
100-200	3.00
300-500	5.50
500-800	7.35
800 +	9.00

The monthly billed amount for the above scenario will be

(100x1.65+100x3+200x5.5+53.80x7.35) = Rs. 1960.43

For more information on case study pls scan

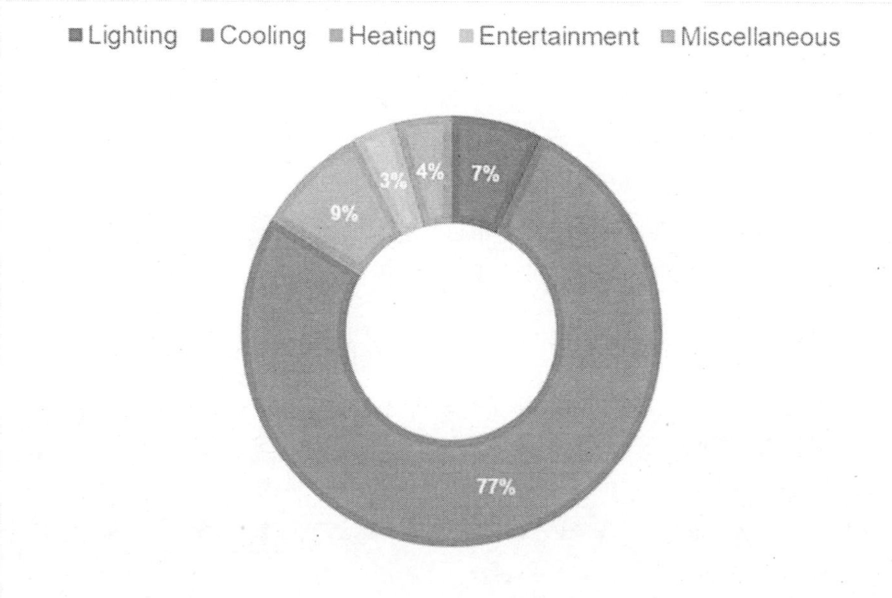

Fig. 7. 4. Energy consumption by different loads of household

Some recommendations for low power consumption in homes

- Take advantage of sunlight and leave lights off during the day
- Replace choke tubelights and CFLs with LEDs
- Turn of the electrical appliances when not in use
- Try to implement sensors for automation of lighting loads and ACs
- Replace fans with low wattage fans such as fans with BLDC motors
- Buy new electronic appliances considering energy star rating
- Refrigerator should be kept almost 0.5 feet away from the wall for optimum cooling
- ACs should be operated in optimum temperature range from 22 to 24 deg Celsius
- Replace CRT TVs with LED TVs

The purpose of the above case study is to verify, monitor, and analyze usage of energy. It also involves submitting a technical report with suggestions for increasing energy efficiency, along with a cost-benefit analysis and an action plan to do so. Every person has a social responsibility to save energy. In this section, we've examined the wattage usage of several gadgets, offered replacement recommendations, and demonstrated net savings.

According to this research, if we utilize energy-efficient equipment, we may save a lot of the power that is now being wasted by our gadgets without affecting their output and use it for other devices. Utilizing energy-efficient technology allows us to conserve energy, lessen power shortages, and control power price inflation.

7.7.Concept of Energy Efficiency

Once the optimization of energy usage and different energy sources is achieved, the next stage is effective operation of the equipment through the best practices as well as judicious technology adoption. Some of the pre-requisites for an energy efficient equipment are:

- Use of efficient semiconductor converters
- Deploy variable speed drivers
- Improve the power factor
- Using single-phase or three-phase semiconductor converters in rural applications
- Enhance the quality of power supply
- Mitigate air leakages in compressors
- Use fuel additive for improvement
- Clean fan blades for exhaust and cooling systems
- Use of proper thermal insulation and double-glazing techniques
- Regular defrosting to reduce energy consumption

The importance of electrical energy efficiency and conservation are:

- The exponential rise in demand of electrical power can be met by conserving power and making electrical systems more efficient
- Electrical power is majorly consumed by various electrical drives
- Sufficient energy can be saved by these efficient and rigid types of electrical drives

7.8.Some examples of energy efficiency and energy conservation

Some of the key techniques to improve energy efficiency and conservation are listed below:

Electrical drives

- Using special type of electrical machines with high power factor
- Deploy a phase advancer on the shaft of the induction motor to supply exciting current to the rotor of the machine
- Promote installation of static capacitors

Household appliances

- Refrigerator should be maintained 37°F - 400°F and freezer sector at 50°F with moisture control
- The item to be place in fridge should be cooled to room temperature
- Use dim light in galleries, lobbies; balconies etc.
- Lamps should be put in those corners of the rooms where they can reflect many light surfaces instead of one.
- Regular cleaning of filter in air conditioning systems and cleaning of condenser conserve energy.

Community level

- Shifting of controllable load from peak demand hours to non-peak demand hours
- Integrate renewable sources of energy
- Installation of photoelectric controls or timers should be used to make sure that outdoor lighting is sufficient during the day.
- Environmentally sustainable transport will promote more mileage less pollution by greenhouse gases.

Industry level

- Regular monitoring and audit of energy consumption.
- Thermal insulation of fuel tanks, ceramic furnaces, etc.
- Old factories should now employ process modification
- Shifting to LEDs for lighting load to reduce power consumption

Summary:

In this chapter basic understanding of energy conservation, energy efficiency and energy auditing are discussed. With suitable examples preliminary and detailed energy auditing and types of energy auditing are presented. A list of equipment required to conduct energy auditing of a building, industry or commercial sector is also presented in the chapter. Advantages of energy conservation and energy auditing, economics analysis of energy auditing are also explained through simple examples. Concept of energy conservation at individual and industrial level are also presented to understand the efficient use of energy.

Short and Long Answer Type Questions

Q1. Explain the importance of energy audit and energy conservation.

Q2. Identify the energy conservation measures that are important in a manufacturing industry.

Q3. Explain the various reasons for increased emphasis on energy conservation.

Q4. What are the various objectives of energy audit? Discuss the different steps for detailed energy audit in an academic institution.

Q5. Write short notes on energy conservation in agricultural sector.

Q6. What different instruments are required for conducting an energy audit of a college building?

Q7. Why an individual should opt for energy auditing of his premises?

Q8. Define and illustrate energy audit, energy efficiency and energy conservation with the help of suitable example

Q9. Briefly discuss the various energy efficient lighting controls in respect of building energy management.

Q10. Discuss some energy conservation measures for house hold and commercial sectors.

Q11. How can the benefits of energy conservation measures be evaluated? Explain

Q12. How can energy audit lead to energy conservation?

Q13 How can energy conservation be promoted in small scale industries?

Problem 1:

Determine the cost associated if you replace all the incandescent lamp and fluorescent/CFL of your home/typical home with the appropriate rating of LEDs lamp. You may assume any appropriate data; in case the data is not available. Also, determine the payback period to recover the cost of replacement.

Assume the following parameters. The requirements of illumination level are: Common Room: 35 Lux; Bed Room: 35 Lux; Reading Room: 250 Lux.

Parameter (s)/Specification (s)	Type of Lamp		
	LED	**Incandescent**	**Fluorescent/CFL**
Luminous Efficiency/Efficacy (lm/W)	70	15	45
Electrical Energy Cost (Rs.) per kWh	6.0		
Average bulb life (Hours)	50,000	1,000	8,500
Cost of Lamp	Rs. 300	Rs. 10	Rs. 100
Operating Hours/day	10	10	10

Assume the missing data, if any.

Problem 2:

Estimate the monthly energy consumption of your home/apartment/typical home according to the following table. Also determine the monthly electricity bill and compare it with your existing bill (If possible).

Description of Load	Rating (W)	Quantity (Numbers)	Total Power (W)	Number of operating Hours	Electrical Energy Consumption (kWh)
Air Conditioner					
Refrigerator					
Freezer					
Oven					
Microwave					
Toaster Oven					
Built-In Dishwasher					
Portable Dishwasher					
Washing Machine					
Dryer					
Coffee Maker					
Vacuum Cleaners					
Blender					

Food Processor					
Rice Cooker					
Juicer					
Coffee Grinder					
Popcorn Machine					
Deep Fryer					
Iron					
Water Heater/Geyser					
Space/Room Heater					
Ceiling Fan					
Incandescent Lamp					
Fluorescent/CFLs					
LEDs Lamp					
Others (if any)					
Others (If any)					
Other (If any)					

Problem 3:

Determine the electrical energy cost for 50000 hours and average cost in 20 years period for the following lamps and fill the blank data in the below table.

Parameter (s)/Specification (s)	Type of Lamp		
	LED	Incandescent	CFL
Wattage (Rating)	10 W	60 W	15 W
Electrical Energy Cost (₹) per kWh (0-200 units) – ₹ 3.00/kWh; (201-400 units) – ₹ 4.50/kWh; (401-800 units) – ₹ 6.50/kWh; (801-1200 units) – ₹ 7.00/kWh; (above 1200 units) – ₹ 8.00/kWh			
Electrical Energy Cost (₹) for 50,000 hours	------	------	------
Average bulb life (Hours)	50,000	1,000	8,500
Cost of lamp	₹ 200	₹ 10	₹ 120
Operating Hours/day	10	10	10
Average Cost (₹) in 20 years period	------	------	------

References and suggested readings

1. B.L. Capehart, W.C. Turner, W.J. Kennedy, "Guide to Energy Management," The Fairmont Press, 4th Edition, 2003.

2. W.J. Kennedy, "How to Conduct an Energy Audit," Proceedings of the 1978 Fall AIIE Conference, Institute of Industrial Engineers, Atlanta, GA, 1978.

3. Energy Audit Handbook, Sustainable Energy Authority of Ireland, Dublin, Ireland, 2015.

4. Energy Conservation Through Effective Energy Utilization, National Bureau of Standards, Washington, USA, 1976.

About the Book

Energy Science and Engineering

This book has been designed for second-year students of undergraduate students in the engineering program. It will prove to be a valuable source for practicing engineers and faculty members. The book is divided into seven units and provides all the necessary information on an introduction to energy systems and renewable energy resources, with a scientific examination of the energy field with an emphasis on alternative energy sources, their technology, and applications.

This book will provide an opportunity to explore society's present energy needs and future energy demands, examine conventional energy sources and systems including fossil fuels and nuclear energy. More focus of this book is on renewable energy sources, sustainability, and the environment. Clean energy technologies and their importance in sustainable development, carbon footprint, energy and environment, trade and research policy, future energy use that can be influenced by the economics, linkage between economic and environmental outcomes are discussed by including the latest available statistical data. This book also covers topics on civil project development for creating energy infrastructure. Concepts of green building and green architecture, energy auditing and energy enterprises with some practical examples are presented in easy language to understand the fundamental concepts. It is important to note that in all the units, dynamic QR codes are provided to collect additional knowledge on the specific topic.